Systematic Studies in the Melastomataceae: *Bellucia*, *Loreya*, and *Macairea*

SUSANNE S. RENNER

MEMOIRS OF THE NEW YORK BOTANICAL GARDEN VOLUME 50

Memoirs of The New York Botanical Garden

The MEMOIRS are published at irregular intervals in issues and volumes of various sizes and are designed to include results of original botanical research by members of the Garden's staff, or by botanists who have collaborated in one or more of the Garden's research programs. Ordinarily only manuscripts of fifty or more typewritten pages will be considered for publication.

Manuscripts should be submitted to the Editor. For further information regarding editorial policy and instructions for the preparation of manuscripts, address the Editor.

Orders for past and forthcoming issues and volumes should be placed with:
Scientific Publications Department
The New York Botanical Garden
Bronx, New York 10458 U.S.A.

THIS PUBLICATION IS PRINTED ON ACID-FREE PAPER.

Systematic Studies in the Melastomataceae: *Bellucia*, *Loreya*, and *Macairea*

Susanne S. Renner

The New York Botanical Garden
Bronx, New York 10458, U.S.A.
Issued: 16 February 1989

MEMOIRS OF THE NEW YORK BOTANICAL GARDEN VOLUME 50

Published by
The New York Botanical Garden
Bronx, New York 10458

International Standard Serial Number 0071-5794

Library of Congress Cataloging-in-Publication Data

Renner, Susanne, 1954–
Systematic studies in the Melastomataceae : Bellucia, Loreya, and Macairea / Susanne S. Renner.

p. cm. — (Memoirs of the New York Botanical Garden, ISSN 0071-5794 ; v. 50)

Summary in Spanish.
Includes index.
Bibliography: p.
ISBN 0-89327-335-X
1. Macairea—Latin America—Classification. 2. Bellucia—Latin America—Classification. 3. Loreya—Latin America—Classification. 4. Botany—Latin America—Classification. I. Title. II. Series.
QK1.N525 vol. 50
[QK495.M514]
581 s—dc19 88-39155
[583′.42] CIP

International Standard Book Number 0-89327-335-X

FEBRUARY 1989 MEMOIRS OF THE NEW YORK BOTANICAL GARDEN **50:** 1–112

Systematic Studies in the Melastomataceae: *Bellucia, Loreya,* and *Macairea*

Susanne S. Renner[1]

Contents

[1] Department of Botany, National Museum of Natural History, Smithsonian Institution, Washington, D.C. 20560. Present address: Botanical Institute, Aarhus University, Nordlandsvej 68, DK-8240 Risskov, Denmark.

Summary

Bellucia is a small neotropical genus of trees and shrubs ranging from Mexico to Bolivia. It has large 5–8-merous flowers. The calyx consists of triangular lobes or is a truncate limb or a calyptra. There are no preformed lines of dehiscence on the calyptra, its breakage occurring in a slightly irregular semicircle. Five of the previously described 27 species are recognized; a new species from Brazil and Bolivia, ***B. beckii*** Renner, is described and illustrated; two new combinations are made, ***B. huberi*** (Wurdack) Renner and ***Miconia cacatin*** (Aublet) Renner.

Loreya is a neotropical genus comprising 13 species of trees, mainly of the upper Amazon valley but ranging from Nicaragua to Bolivia. Careful consideration of its relationships with the allied genera *Henriettella, Henriettea, Myriaspora,* and *Bellucia* shows that *Loreya* is a distinct taxon defined by its characteristic stamen morphology and the usually strongly plinerved, ovate, large, and—when dry dark brown leaves. Four new species from Colombia, Peru, Brazil, and Bolivia are described: ***Loreya klugii*** Renner, ***L. riparia*** Renner, ***L. wurdackiana*** Renner, and ***L. gracilis*** Renner. Two new combinations are proposed, ***L. subrotundifolia*** (Wurdack) Renner and ***Henriettea mucronata*** (Gleason) Renner.

Macairea, a neotropical genus of 22 species of shrubs and subshrubs, is characterized by unique dorsi-basal connective appendages and glandular hairs on the vegetative and fertile parts of the plants. Most species have terminal, many-flowered, cymose inflorescences with 4-merous, purple flowers with white centers; in two species, the inflorescences are borne in the axils of the uppermost leaves and in one, the flowers are subsessile. The seeds are cochleate or subcochleate and in a few species polyhedric, with a clearly defined circular subterminal hilar region. These seeds place *Macairea* in the Tibouchineae, where it is closely related to *Tibouchina. Macairea* species require ample light and are adapted to acidic and/or waterlogged soils; all occur in savanna habitats. *Macairea* is most diverse in the Orinoco-Rio Negro drainage and in the Gran Sabana (Venezuela); in these two areas 17 of the 22 species are represented. Seven species are found exclusively at higher elevations on the sandstone mountains of Guayana. Twenty of the previously described 43 species are recognized and two new ones, ***Macairea philipsonii*** Renner and ***M. cuieirasii*** Renner, are described; two species are illustrated.

Sumário

Bellucia é um gênero neotropical, de poucas espécies, de plantas arbóreas e arbustivas, occorendo desde o México até a Bolívia. As flores variam de pentâmeras a octâmeras. O cálice pode ser caliptriforme ou truncado ou apresentar lobos triangulares. No cálice caliptriforme não se observa linha de ruptura; a deiscência ocorre por um semicírculo irregular. Somente cinco das 27 espécies descritas são reconhecidas no presente trabalho; uma nova espécie ocorrente no Brasil e na Bolívia é descrita e ilustrada: ***B. beckii***; e duas novas combinações são propostas: a transferência de *B. cacatin* Aublet para o gênero *Miconia* e a de *Loreya huberi* Wurdack para *Bellucia.*

Loreya é um gênero neotropical, compreendendo 13 espécies arbóreas e ocorrendo desde a Nicaragua até a Bolívia, principalmente na região oeste da bacia amazônica. Considerações criteriosas sobre suas relações com os gêneros afins *Henriettella, Henriettea, Bellucia* e *Myriaspora* evidenciam que *Loreya* é um taxon bem definido, distinto pela morfologia estaminal, o hábito arbóreo e pelas folhas que, em sua maioria, apresentam padrão de nervação acródromo suprabasal, forma ovada e, quando secas, cor castanho-escura. Quatro espécies novas são descritas para a Colombia, Perú, Brasil e Bolívia: *L. klugii* Renner, *L. riparia* Renner, *L. wurdackiana* Renner e *L. gracilis* Renner. Duas novas combinações são propostas: *L. subrotundifolia* (Wurdack) Renner e a transferência de *L. mucronata* Gleason para o gênero *Henriettea.*

O gênero *Macairea* é neotropical e constituido de 22 espécies arbustivas e subarbustivas. Caracteriza-se por apresentar conectivo com apêndice dorso-basal e tricomas glandulares tantos nas partes vegetativas como florais. A maioria das espécies tem inflorescências cimosas, terminais, multifloras, com flores tetrâmeras, roxas, mas brancas no centro; em duas espécies as inflorescências originam-se nas axilas das folhas superiores e em uma outra espécie as flores são subsésseis. As sementes são cocleares ou subcocleares e em poucas espécies, poliédricas, com região hilar subterminal, circular. Estas caracteristicas morfológicas das sementes permitem posicionar *Macairea* na tribo Tibouchineae, próximo de *Tibouchina.*

As espécies de *Macairea* exigem muita luz e estão adaptadas a solos ácidos e/ou alagados; todas ocorrem em savanas. O maior número das espécies ocorre na região banhada pelos rios Orinoco e Rio Negro e na Gran Sabana (Venezuela): nestas duas áreas encontram-se 17 das 22 espécies de *Macairea.* Sete espécies são exclusivas das regiões montanhosas de Guayana. Das 43 espécies descritas somente 20 são reconhecidas no presente estudo e duas novas espécies são descritas: ***M. philipsonii*** Renner e ***M. cuieirasii*** Renner; duas outras são ilustradas.

General Introduction

In the neotropics, the Melastomataceae and Memecylaceae, two families often united, comprise over 3000 species in 106 genera. Most Melastomataceae are characterized by poricidal anthers and by a pronounced foliar venation consisting of the mid-vein and one to four pairs of subparallel longitudinal veins. The main dichotomy in the Melastomataceae is between the capsular-fruited and the berry-fruited genera. The genera are arranged into tribes by a combination of fruit, seed, and stamen characters.

Though substantial numbers of Melastomataceae species have been or are being dealt with in modern floristic treatments (Wurdack, 1962, 1973, 1980a, and in prep.; Almeda, in prep.), there are few monographic studies. I became interested in the family during field work on melastome reproductive biology and it seemed appropriate to continue working with a group I knew well. *Bellucia,* an apparently well-defined genus, was of particular interest to me because I had studied the breeding systems of several of its species and because, in spite of its small size, it offered much variation in morphology and habitats occupied. *Bellucia* is in the berry-fruited tribe Miconieae, the largest of the family. Most of its species occur in lowland rain forest and are characterized by a calyptrate bud cover, a trait which has arisen many times independently in the Melastomataceae (e.g., *Centronia, Conostegia, Graffenrieda, Mecranium*). As it turned out, *Bellucia* is monophyletic, but *Loreya,* a genus traditionally seen as very close or synonymous, and which I therefore initially regarded as a sister group to *Bellucia,* is likely paraphyletic if *Bellucia* is excluded. These problems naturally led to a revision of *Loreya.* I am maintaining the two genera (and transferring species in both directions, as well as describing several new ones), even though I recognize the problems of keeping possibly paraphyletic groups. My reasons for maintaining *Bellucia* as distinct from *Loreya* are given in the present treatments.

While working on the floral biology of some Amazonian species, I found that many were capable of producing seeds asexually. The first apparently predominantly agamospermous species I encountered was *Macairea theresiae.* Later in Washington, John Wurdack told me that H. A. Gleason had given him the choice of monographing either *Brachyotum* or *Macairea* (a genus in which Gleason took particular interest) for his dissertation. Upon throwing a coin, Wurdack ended up doing *Brachyotum. Macairea* is a capsular-fruited genus traditionally, and in my opinion correctly, placed in the Tibouchineae, one of five capsular-fruited neotropical tribes. It is a monophyletic genus and I consider it to be close to *Tibouchina. Macairea* has a clear center of diversity in the Guayana phytogeographical province. Potentially agamospermous plants occurring in isolated and topographically diverse habitats are of special interest for understanding evolution and speciation in higher plants. In addition, *Macairea* is one of the capsular-fruited Melastomataceae among the marker species characterizing different types of neotropical savannas.

Notes on Measurements, Abbreviations, and the Citation of Photographs

Measurements are taken from dried material; petal and fruit colors refer to living material. For *Macairea,* the length of the longer, antesepalous stamens is given before the length of the shorter, antepetalous stamens. Morphological terms used in the key and descriptions are discussed in the relevant general chapters and/or illustrated (e.g., hair types, seed surfaces).

Photographs have been cited by negative number where known, preceded by the herbarium acronym or photographer's name (H. A. Gleason). Thus "F photo neg. . . ." refers to a negative from the Field Museum of Natural History, J. F. Macbride type photo collection. Herbaria acronyms follow Holmgren et al. (1981); all herbaria from which material was consulted are listed in the "Acknowledgments" section, p. 93.

Taxonomic Revision of *Bellucia* (Miconieae)

Introduction

Bellucia is a genus of neotropical trees and shrubs in the Melastomataceae comprising seven species. It ranges from southern Mexico to northern Bolivia and east through the Guianas into Maranhão in Brazil. These plants are collected frequently because of their abundance in a wide range of habitats, and because they flower and fruit conspicuously and more or less year-round. My interest in *Bellucia* developed in connection with field work on the reproductive biology of Melastomataceae carried out in Brazilian Amazonia and I have been fortunate to study all the species in their natural habitats, except for two species recently discovered. The need for revision has long been apparent; *Bellucia* was last treated as a whole by Cogniaux (1891a) who saw about 50 collections. The ca. 1000 collections which I could study have shown that of the 27 species described, only five can be recognized; also, I have described one new species and transferred another from *Loreya* into *Bellucia.*

Taxonomic History

The first known published description and illustration of a member of *Bellucia* is by Plukenet (1692). The plate shows a single leaf of a plant described as "Arbor Americ., latiori & acuminato folio trinervi, utrinq. glabro, & margine leviter crenato. Mespilboom Belgis sirinamensibus nuncupatur." The leaf from which this plate was made is now in the Sloane herbarium (vol. 95, fol. 78). It is typical of the genus in size and shape, the glabrous smooth surface, and the fine reticulate venation of the lower leaf surface. The vernacular name given by Plukenet, "mespilboom" (or "mespelboom"), is still used for the single species of *Bellucia, B. grossularioides,* occurring in the Guianas; it is, however, also used for other arborescent Melastomataceae. There is also a fertile specimen in the Vaillant herbarium (P) with a label repeating Plukenet's description. This may be a duplicate Plukenet sent to Vaillant.

The first validly published name applicable to *Bellucia* is *Melastoma grossularioides* Linnaeus. Linnaeus (1753: 390) described the plant with an original diagnostic phrase name and with reference to three elements: the Plukenet illustration, a synonym for Plumier (1755–1760), and the name "Bartsch."

Johann Bartsch (1709–1738), a German collecting in Surinam for Boerhaave, corresponded with Linnaeus in 1737 (Linn. Corr. **1:** 426) but no reference to *Bellucia* is made in any of the extant letters. There seems to be no Bartsch specimen either among the Boerhaave plants (Herb. Sloane; Dandy, 1938; C. Jarvis, pers. comm.), or in the Linnaean herbaria (LINN, OXF, S). The Plumier element is not a *Bellucia*; in Linnaeus' own annotated copy of *Species Plantarum,* he crossed out the Plumier synonym, and this was omitted from the second edition of *Species Plantarum* (1762: 558). Plumier's illustration shows *Clidemia guadalupensis* (DC.) Grisebach (fide Urban, 1920), which has the purple hairy fruits to which Plumier's manuscript name "*grossularioides*" (gooseberry-like) refers. By elimination, *Melastoma grossularioides* is typified by the Plukenet plate.

Aublet (1775) also collected the species and described it as *Blakea quinquenervia.* His decision to place the species in P. Browne's genus *Blakea* was followed by Linnaeus' son (1781), Jussieu (1789), David Don (1823), and de Candolle (1828a). However, Necker (1790) held that Aublet's species should be placed in a separate genus and proposed the unitary designation *Bellucia,* honoring Toma Bellucci (?–1671), head of the Pisa Botanical Garden and author of *Plantarum index horti Pisani* (1662). Necker's name was validly published as a generic name by Rafinesque (1838) and—following a proposal by Rehder (Rehder et al., 1935)—conserved over the name *Apatitia* published by Desvaux (*in* Hamilton, 1825). About 35 years before Desvaux, Gmelin (1791) had already validly published *Webera quinquenervia*; however, this was a later homonym of Schreber's *Webera* (Rubiaceae) and thus had to be rejected. Both Necker and Desvaux refer to the fact that *Bellucia* is similar to *Blakea*: Necker coined the French name "Fausse-Blakée" meaning "pseudo-Blakea" and Desvaux the name *Apatitia blakeoides*, "Apatitia" from a Greek word meaning "to deceive."

During the course of his work on the Melastomataceae in London (Triana, 1871, introduction), Triana examined the type of *Blakea quinquenervia* Aublet and the leaf of *Melastoma grossularioides* L. illustrated on the Plukenet plate and recognized that they represented the same species. The leaf is determined in Triana's hand as *Bellucia aubletii* Naudin (C. Jarvis, pers. comm.), an illegitimate name for *Blakea quinquenervia* Aublet. In this final treatment (1871), Triana synonymized both these names under *Bellucia grossularioides*.

Naudin (1850c) described five new species of *Bellucia* bringing the total to six. After having seen *Bellucia* in its natural habitat, Seeman (1854) commented "I do not think that the five species of *Bellucia* proposed by Naudin will be found to be good species; to me they appear mere varieties." Nevertheless, one of Naudin's species, *B. pentamera*, based on a collection made by Claude Gay in Peru, was a good species; it is in fact the most widespread and most frequently collected one. Unfortunately, Naudin's name was hardly ever used probably because there were no duplicates of the Gay collection in herbaria other than Paris.

A beautiful colored engraving of *Bellucia pentamera* accompanied by detailed descriptions of habit, habitat, and morphological details was made for Karsten's *Flora Columbiae* (1861, pl. 87). Karsten (1859, 1861) felt that the difference in calyx structure between the then known *Bellucia* species warranted generic status, and he proposed *Axinanthera* as a name for the segregate genus.

Cogniaux (1888, 1891a, 1891b, 1908) reunited *Axinanthera* with *Bellucia*, recognized seven of the earlier species, and described five new ones, only one of which is here recognized. Four additional species have been described in this century (Pilger, 1905; Wurdack, 1980b, 1982) and the two widespread species, *Bellucia grossularioides* and *B. pentamera* (as *B. axinanthera* or *B. costaricensis*), have been treated in the floras of Guadeloupe and Martinique (Fournet, 1978), Guiana, Surinam, and Panama (Gleason, 1932, 1935, and 1958, respectively), Peru (Macbride, 1941), Costa Rica, Mexico, and Guatemala (Standley, 1938, 1961, and Standley & Williams, 1963, respectively), the Province of Cundinamarca in Colombia (Uribe, 1972), Venezuela and Ecuador (Wurdack, 1973 and 1980a), and Barro Colorado Island, Panama (Croat, 1978).

Whereas the majority of workers considered *Bellucia* very close to *Blakea*, Baillon (1877) thought that *Bellucia* was really the same as *Loreya* De Candolle. *Loreya*, at the time, comprised six morphologically diverse species of which two, *L. spruceana* Bentham ex Triana and *L. mespiloides* Miquel, according to Baillon resemble *Bellucia* in their floral structure. Macbride (1941) went so far as to synonymize *Loreya* under *Bellucia*.

Vegetative Anatomy

The wood anatomy of the neotropical Melastomataceae has been studied in Utrecht (Koek-Noorman et al., 1979; Vliet et al., 1981; Welle & Koek-Noorman, 1978, 1981; Welle & Mennega, 1977).

These workers found that within the Miconieae, wood anatomical characters are rarely diagnostic at the generic level. However, the size and shape of the intervascular pits and the size of the vessel ray pits sometimes have taxonomic value (Welle & Koek-Noorman, 1981). Here a brief summary is provided of the most important wood features of *Bellucia* based on Welle and Koek-Noorman's analysis of *B. acutata*, *B. pentamera* (as *B. axinanthera*), and *B. grossularioides*. (The samples *Lindeman 6462a* and *Krukoff 5580*, identified as *B. imperialis*, are in fact *B. grossularioides* and *B. pentamera*, respectively.)

Growth rings scarce, faint, or absent. Vessels diffuse, solitary, in radial multiples and irregular clusters of 2–8, 4–17 (1–27) per mm^2. Intervascular pits alternate, vestured, round and only sporadically polygonal or oval, 4–8 μm. Vessel-ray pits round, oval, and oblong, 5–20 × 5 μm.

Fibers septate, gelatinous fibers scarce to abundant. Fiber/vessel ratio 1.20–1.51. Rays uniseriate. Parenchyma scanty paratracheal.

Bellucia wood is relatively light (0.50–0.75 g/cm^3; Détienne & Jacquet, 1983), soft, and brittle. Its color is tan to rosy brown.

The leaf anatomy of most genera in the Miconieae has been thoroughly investigated by Gottschall (1900; for summaries, see Solereder, 1908, and Metcalfe & Chalk, 1979). He studied *Bellucia grossularioides* and *B. dichotoma* and noted that the leaf epidermis cells have a particularly thick cuticle which is smooth on the upper surface and papillose on the lower surface. An SEM survey of the lower leaf surface revealed the papillae to be characteristic of all species of *Bellucia* (Fig. 1G, H). These papillae cause the glaucous appearance of the lower leaf surface.

In leaf clearings, I found clustered calcium oxalate crystals (druses), styloids (i.e., elongated prismatic crystals), and crystal sand in the leaf spongy parenchyma. The presence of crystals is typical of *Bellucia* and related genera (Baas, 1981; pers. obs.). The secondary and tertiary veins are surrounded by a sclerified sheath, and the tracheids have spirally reinforced walls in the parenchyma. Stomata occur only on the lower leaf surface.

MORPHOLOGY

Vegetative Structures

The species of *Bellucia* are usually medium-sized trees; three species are shrubs. Sufficient data on growth form from field observations and herbarium labels are available to show that the arborescent species differ in the heights they attain under normal conditions. *Bellucia grossularioides* grows to a height of 25 m and may reach 30 m with a DBH of 30–60 cm, occasionally 100 cm; *B. pentamera* and *B. dichotoma* hardly exceed 15 and 12 m in height, respectively (with a DBH of 20–30 cm). *Bellucia aequiloba* is a tree to 20 m in height.

Bellucia trunks are cylindrical; the branches depart at 60° angles. The growth form of *B. grossularioides* is monopodial with rhythmic branching, the branches being orthotropic and sympodial. This approaches Scarrone's model in the system of growth forms by Hallé et al. (1978; Cremers, 1983).

Bellucia trees for the most part grow exposed and have round crowns; however, growing as part of a cohort of trees in a forest gap results in unbranched trunks with an umbrella-type crown.

The bark is about 1 cm thick (in the arborescent species) and its appearance is species-specific. The inner bark of all species is orange to red, the outer bark is brown or sometimes gray (due to the presence of *Chiodecton* lichens) and rough with shallow fissures or cracks and pustules. The three shrubby species have smooth bark.

The leaves of *Bellucia* are opposite, estipulate, coriaceous, and generally large, i.e., 15–26(–42) × 12–18(–39) cm. The venation is 3–5-plinerved, with the inner pair of lateral primaries departing from the mid-vein 0.5–3 cm above the blade base. When fresh, mature leaves are dark green and glossy above and dull green and glaucous below. Juvenile leaves are slightly lighter in color and sometimes have a serrulate or denticulate margin. Upon drying, the leaves become olive to dark brown and the glaucescence on the lower surface becomes more pronounced. Among the neotropical Melastomataceae, *Bellucia* leaves are distinct in their leathery texture, broadly ovate shape (with the exception of *B. acutata* which has oblong leaves), smooth upper surface, and glaucous lower surface. They can only be confused with those of *Henriettella prancei* Wurdack. Leaves of *Bellucia grossularioides* cannot be distinguished from those of *B. pentamera.*

The pubescence of the lower leaf surface consists of thin (apically sometimes slightly curled) hairs each comprising a single row of cells (Gottschall, 1900, fig. 11; Metcalfe & Chalk, 1979, fig. 5.2.1; Fig. 1G). Where these smooth trichomes reach a high density, the surface feels felty or even velvety to the touch.

The petioles are dark brown and channeled adaxially.

Reproductive Structures

The *inflorescences* are cymes or rarely single flowers; in most species they are borne in the axils of still existing leaves. In *Bellucia dichotoma* and *B. aequiloba,* inflorescences develop exclusively on the old wood, which means that a number of years pass between the abscission of the supporting leaf and the further development of the inflorescence initiated as a bud in its

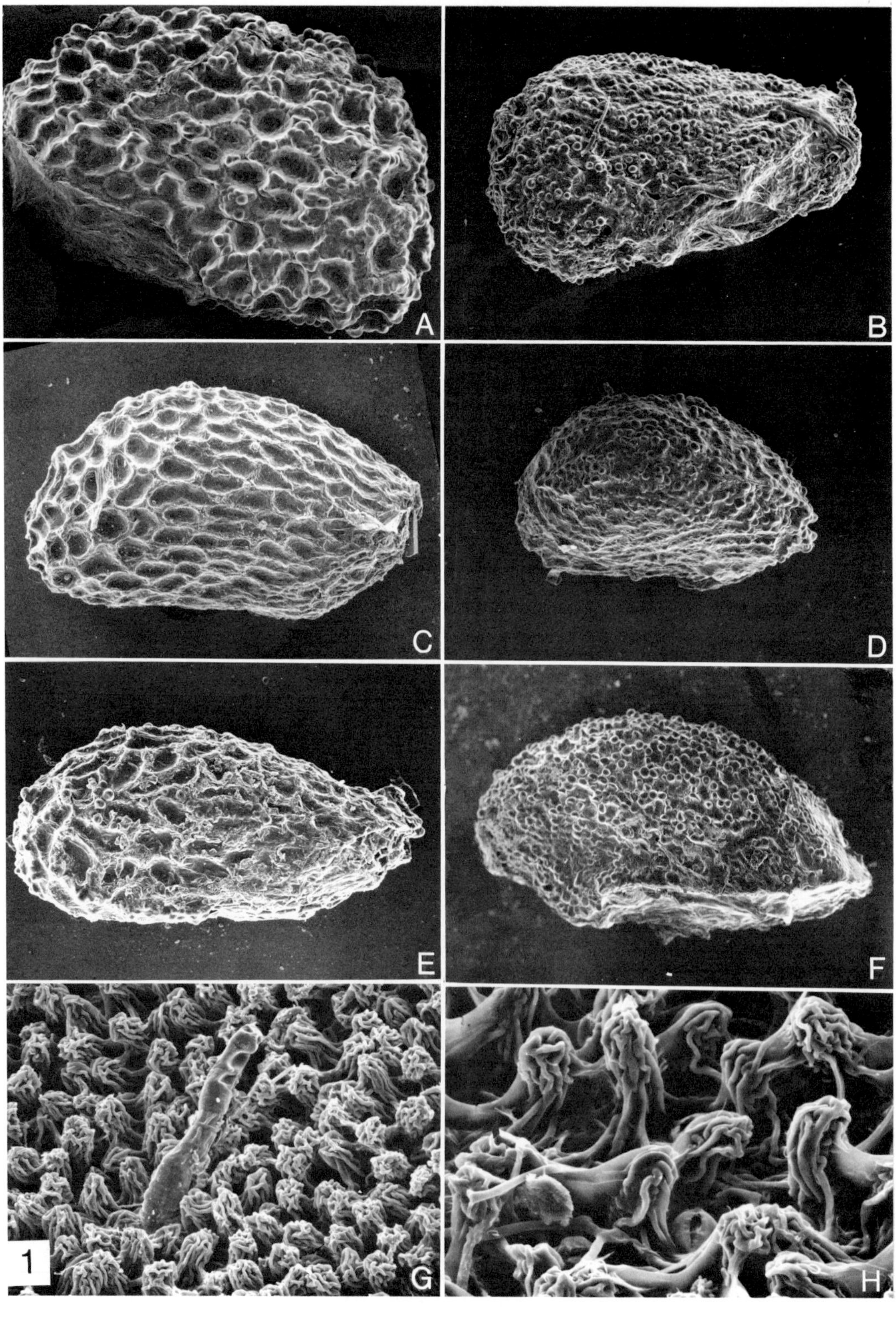
A
B
C
D
E
F
1
G
H

axil. The cauline inflorescences are 3–4-branched with eight to 16 flowers, whereas the axillary inflorescences are mostly 1-branched with two flowers, perhaps due to the limited space in the axils of existing leaves.

In two species, there is some variability in the extent of inflorescence-branching. In *Bellucia grossularioides,* the degree of branching is correlated with geographic distribution. The more branched inflorescences are found in the vicinity of Villavicencio (Colombia), the upper Rio Negro region (Venezuela/Brazil), the state of Bolívar (Venezuela), and along the Rio Madeira in the southwestern part of the state of Amazonas (Brazil). However, I have observed trees with branched inflorescences growing next to individuals with unbranched ones. The variability in the degree of branching does not seem to warrant taxonomic recognition as it is not correlated with any other morphological characters.

Bellucia flowers have pedicels bearing a pair of scale-like, caducous bracts on the lower third. The hypanthium is completely fused with the ovary wall, subglobose, and covered with lenticels that become more visible as the fruit matures. The torus is not marked on the outside by a pronounced hypanthium constriction (in contrast to species in some related genera). The calyx consists of triangular lobes or is a subtruncate limb or a calyptra. In the species with triangular calyx lobes, the lines of separation are clearly visible even in very young buds, whereas in the species with a calyptra, the buds are completely smooth. At anthesis, the calyptra dehisces in a slightly irregular semicircle and peels off as a hyaline membrane that persists on developing fruits (Fig. 11A). Petals of the calyptrate species have a smooth exterior surface; petals in the species with triangular sepal lobes (*B. pentamera* and *B. aequiloba*) have a fleshy keel on the outside, which is an inverse impression of the fusion lines of the sepals.

The flowers are usually large, with a diameter to 8 cm. In most species, the number of flower parts is not completely fixed; for example, an individual of *Bellucia grossularioides* may have the number of petals varying from 7 to 9, the number of stamens from 14 to 18, and that of ovary locules from 12 to 15. (The ranges are given in the species descriptions.) Except for differences in size, the morphology of the stamens, and styles is homogeneous among the species. Flowers of all species emit a pleasant perfume. In *Bellucia grossularioides* and *B. acutata,* the odor is produced by the petals, the upper part of the filaments, the connective tissue, the pollen, the upper section of the style, and the stigma (determined by neutral red staining; method from Vogel, 1963).

The petals are fleshy and completely white or white inside and pink-flushed outside. They become brownish on the second day after opening. The petals and stamens are generally shed on the morning of the third day. Upon opening of the flower, the stamens form a closely packed ring around the style; subsequently, the style bends to the lower (closest to the ground) side of the flower and finally comes to lie between two stamens with the stigma protruding from the staminal ring (Fig. 5D). The stigma is large (about 2–3 mm diam.) and sticky and has 8–18 lobes.

Bellucia fruits are many-seeded (e.g., 3200 seeds in *B. dichotoma*), pale green to yellow berries with yellowish pulp. They taste sweet to slightly sour and are somewhat astringent. The *seeds* are 0.5–1 mm long and roughly ovoid with a small, subterminal hilum (Fig. 1A–F). The anticlinal walls of the testa cells are undulate or wavy, the periclinal walls bulge outwards. Sometimes, there is an irregular pattern of shallow grooves formed by larger groups of cells (Fig. 1A, C, E).

Chemistry

Little is known about *Bellucia* chemically. A survey of leaf material of *B. dichotoma* for the occurrence of alkaloids revealed only a large number of flavones and organic acids (H. Wagner, Institute for Pharmaceutical Biology, Munich, pers. comm.). *Bellucia pentamera* and *B. grossularioides* accumulate aluminum (Chenery, 1948, and in a letter to J. J. Wurdack where the

←

FIG. 1. Scanning electron micrographs of seeds and foliar surfaces in *Bellucia.* **A.** *B. grossularioides* (×65). **B.** *B. acutata* (×60). **C.** *B. pentamera* (×80). **D.** *B. dichotoma* (×60). **E.** *B. aequiloba* (×60). **F.** *B. beckii* (×40). **G, H.** Lower leaf surfaces. **G.** *B. aequiloba* (×640). **H.** *B. dichotoma* (×250).

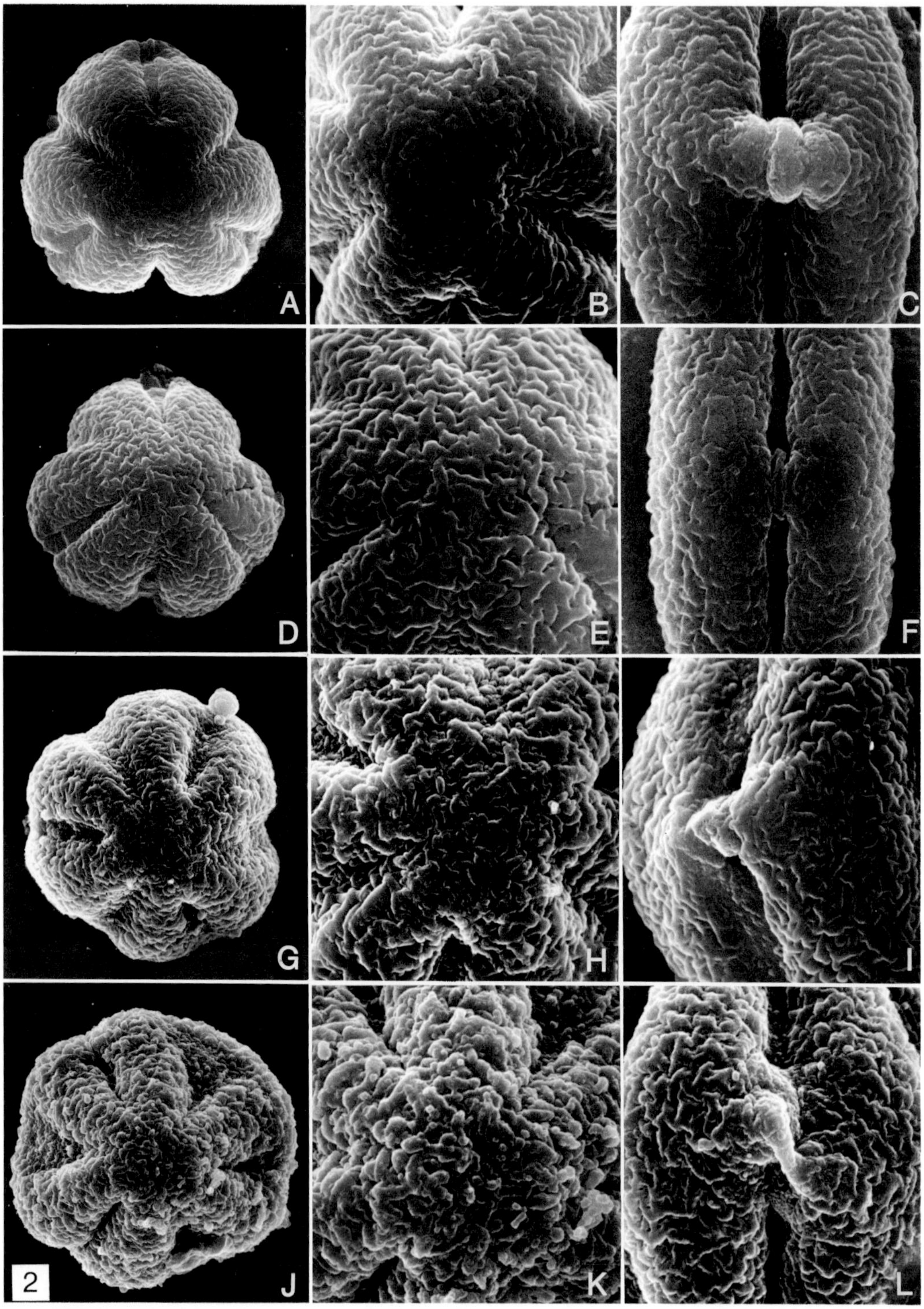
A
B
C
D
E
F
G
H
I
2
J
K
L

names of the species analyzed are given). *Bellucia* fruits contain appreciable amounts of pectin and tannin (pers. obs.).

Pollen Morphology

I acetolyzed pollen according to standard procedures outlined in Erdtman (1960). Samples for scanning electron microscopy (SEM) were first carbon-coated and then sputter-coated with gold-paladium and examined with a Hitachi S-570. Slides for light microscopy were made with glycerin jelly and sealed with wax; measurements are based on 7–10 grains. Only one pollen preparation per species was made. The terms used here to describe pollen grain morphology are those of Erdtmann (1971) and Patel et al. (1985). Light microscope slides of all samples are deposited at the Palynological Laboratory, Department of Botany, Smithsonian Institution.

Pollen grains of all species I examined with SEM were essentially similar in their exine sculpture (Fig. 2C, F, I, L). The grains are shed as monads, are radially symmetrical, and isopolar, their surface is rugulate. The grains have 3-colporate apertures alternating with three subsidiary colpi. Additionally, there are bridges of ektexine (Fig. 2C, F, I, L) over the endoapertures. Grain sizes for the six species are (polar measurements given first followed by equatorial measurements): *Bellucia acutata* (9–)10.2(–11) × (7–)7.2(–8) μm, *B. aequiloba* 10 × (7–)7.2(–8) μm, *B. beckii* (8–)8.3(–9) × 6 μm, *B. dichotoma* 10 × 9 μm, *B. grossularioides* (9–)9.4(–10) × (6–)7.1(–8) μm, *B. pentamera* (8–)9.5(–10) × (7–)7.1(–8) μm.

Pollination and Dispersal

As I have already given a detailed account of the reproductive biology of *Bellucia* (Renner, 1986/1987), only a brief summary is presented here.

Bellucia flowers are exclusively visited and pollinated by female bees of many different bee families. Adaptations to bee pollination are the diurnal anthesis, the flower odor, the color (white petals with the outside sometimes pink-flushed, white and yellow stamens), the style movement, which brings the sticky stigma into a position suitable for contact with the abdomen of a bee, and the poricidal anthers themselves, which can only be exploited by bees. The bees' reward is the copious pollen collected as food for their larvae (details in Renner, 1984a, 1986/1987).

Bellucia acutata, B. grossularioides, and *B. dichotoma* are self-incompatible (Renner, in press). The wet stigmatic surface and the bi-nucleate pollen (Tobe & Raven, 1983) indicate a gametophytic mechanism of self-incompatibility (Nettancourt, 1977).

Bellucia fruits are eaten by a wide range of animals: humans, monkeys (M. van Roosmalen, pers. comm.), bats (S. Marquez, pers. comm.), marsupials (Atramentowicz, 1982), birds (pers. obs.), and, once they have fallen to the ground, by turtles (M. van Roosmalen and D. Moskowitz, pers. comm.), tapirs (Huber, 1909: 197; label data), and ants (pers. obs.). Judging by the considerable size, agreeable smell, and greenish-yellow color of the fruits, monkeys are probably the original dispersal agents; perhaps bats take over in more disturbed habitats. Some of the numerous seeds thus reach the open habitats required for the establishment of the seedlings (see Distribution and Ecology).

Large black ants of the genus *Crematogaster* are frequently found on *Bellucia* trunks and branches and they sometimes carry parts of flowers and fruits lying under the trees to their nests. *Bellucia* fruits are commonly preyed upon by Diptera larvae (Carreira, 1976; pers. obs.).

Chromosomes and Hybridization

Solt and Wurdack (1980) reported a chromosome number of $n = 20$ (21) for *Bellucia pentamera* (as *B. axinanthera*).

Artificial crossings between *Bellucia grossularioides* and *B. dichotoma,* performed near Manaus, yielded fruits with viable seeds (Renner,

←

FIG. 2. Scanning electron micrographs of pollen grains in *Bellucia.* For each species the left micrograph shows a polar view (×2500), the micrograph in the middle of a polar view (×5000), and the one to the right a lateral view (×5000). **A.** *B. acutata.* **D.** *B. grossularioides.* **G.** *B. dichotoma.* **J.** *B. pentamera..*

1986/1987), which indicates that the two species have similar genomes. Fertile adult trees with characters intermediate between these two species occur frequently along roadsides in the vicinity of Manaus; in herbarium material from Bolivia, I found a suspected hybrid between *Bellucia aequiloba* × *B. pentamera* and in material from Guadeloupe, Venezuela, Brazil, and Peru, putative hybrids between *B. grossularioides* × *B. pentamera.* However, no hybridization appears to occur between the arborescent and the shrubby species. The morphological characters of the hybrids are discussed under *B. grossularioides, B. pentamera,* and *B. aequiloba.*

Distribution and Ecology

Bellucia is distributed from Veracruz, Mexico, south through Central America to northern Bolivia and east through the Guianas to Bahía and Maranhão, Brazil (Fig. 3). This corresponds to a latitudinal distribution from 20°N to 17°S.

The center of species diversity, as represented in Figure 3, is the upper Rio Madeira region at the southwestern periphery of the Amazon basin where five of the seven species occur. This is unusual for the Melastomataceae, which is most diverse in the Andes and on the Guayanan and Brazilian shields.

Two species, *Bellucia pentamera* and *B. grossularioides,* have been introduced and become naturalized in various southern Caribbean islands (Grisebach, 1864). *Bellucia pentamera* has also been introduced and is now regenerating naturally in Africa, Malaysia, Borneo, Sumatra, and Java (Helten, 1923; Steenis, 1975; Stone, 1972). Its introduction to Surinam was apparently unsuccessful (see discussion under *B. pentamera*).

The species of *Bellucia* occur primarily in wet tropical lowland forest. Two Andean cordillera species, *Bellucia aequiloba* and *B. pentamera,* occur at altitudes up to ca. 1600 m. With respect to soil and water requirements, the widespread species are tolerant of a range from sandy, well drained to waterlogged (at least for part of the year) soils. *Bellucia pentamera* grows in the very humid climate of the Chocó region of Colombia and also in the relatively seasonal climate of Humaitá (mid-Madeira river) and southeastern Pará in Brazil. *Bellucia aequiloba* apparently requires richer clay loam; *B. dichotoma* seems to tolerate the poorest soils. The latter, which has endomycorrhiza (St. John, 1980), sometimes grows in the low, open forests ("campinarana") on white sand soils in the Rio Negro region; the possible presence of mycorrhizae in any of the other species has not been investigated.

Logging, road-building, and pasturing have provided excellent habitats for the light-demanding *Bellucia* species. In central Amazonia, *B. dichotoma* and *B. grossularioides* are amongst the most important pioneer species in terms of numbers of individuals per area (Renner, unpubl. data). In extra central-Amazonian regions, newly deforested land is often invaded by *B. pentamera.* The three species start flowering after 4–5 years.

The shrubby species, *Bellucia huberi, B. acutata,* and *B. beckii,* grow in open grasslands. Such savannas may occur on rocky soils or on deep soils. They may sometimes be regularly inundated and subjected to fire. The three species replace each other from north to south (Fig. 8); *Bellucia huberi* occurs in savannas in Territorio Federal Amazonas, Venezuela, *B. acutata* is characteristic of the savannas of the lower Rios Branco, Negro, and Madeira, and *B. beckii* grows in north Bolivian grasslands and adjacent cerrado areas of Mato Grosso and southern Pará, Brazil.

Uses

The fruits of all species of *Bellucia* are edible, the larger fruited (arborescent) species have been included among the indigenous potential fruit trees of South America (Cavalcante, 1979; Duarte et al., 1979; Lemée, 1956; Pittier, 1926; Romero-Castañeda, 1961), and *Bellucia pentamera* was introduced into the paleotropics for this purpose. The fruits have also been fed to domesticated animals, for instance donkeys and hogs (label data).

The red inner bark of *Bellucia dichotoma* and *B. grossularioides* is used to waterproof wooden implements, paddles, and drinking vessels. Ropes are soaked in a decoction of the bark to enhance durability. Near Santarém (Brazil), the bark of *B. dichotoma* and *B. grossularioides* is thought to have medicinal properties (pers. obs.), and the Waunana indians in the Chocó area (Colombia) use a decoction of the leaves of *B. pentamera* as a remedy for eye problems (Forero Pinto, 1980).

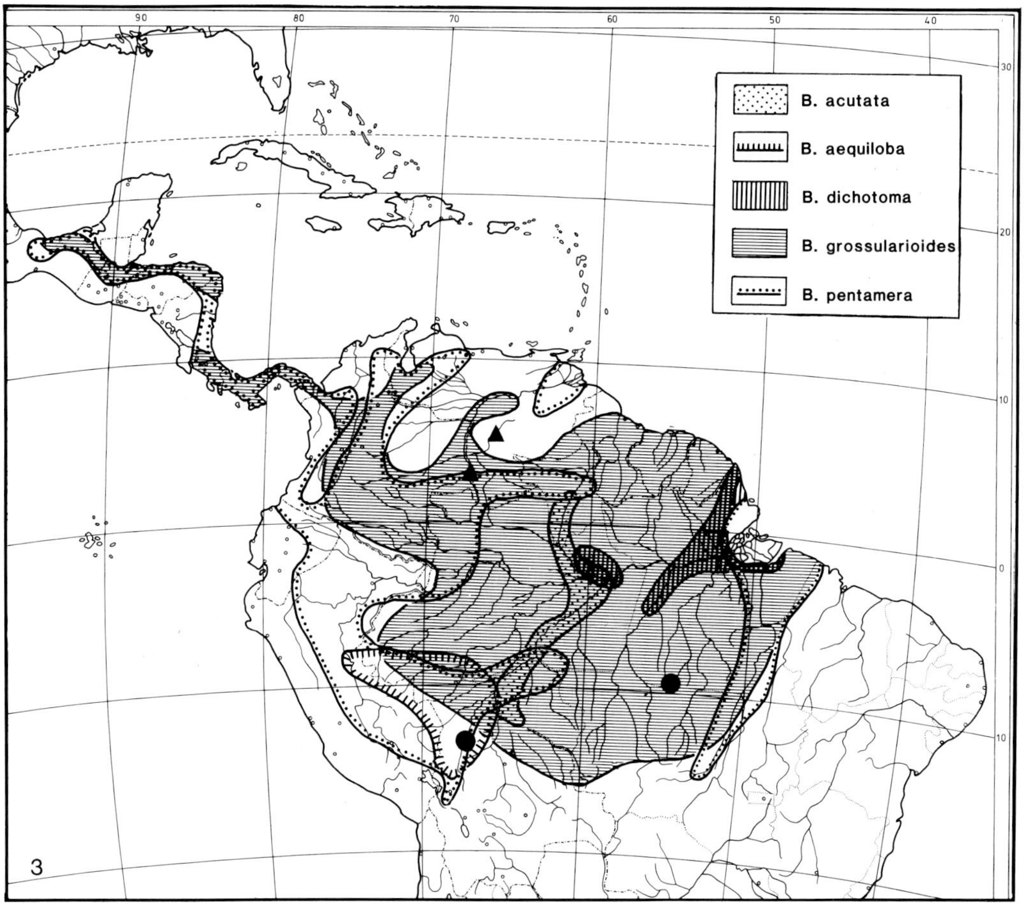

FIG. 3. Distribution of the species of *Bellucia*. *B. huberi* (▲), *B. beckii* (●).

INTER- AND INTRAGENERIC RELATIONSHIPS

The floral similarities of *Bellucia* and *Blakea* led several classical authors to the assumption that the two genera were closely related (see TAXONOMIC HISTORY). *Blakea* (which together with *Topobea* forms the Blakeeae) is characterized by persistent bracts and a particular leaf venation with numerous, fine, linear, lateral secondary veins running so close to each other that they almost give the effect of striations. This type of venation is absent in *Bellucia* and all other Miconieae. *Blakea* seeds are smooth and wedge-shaped or angular (Renner, pers. obs.), not roughly ovoid with a small subterminal hilum and bulging periclinal cell walls as those of *Bellucia*. In addition, some species of *Blakea* have stipular structures, and most are epiphytes with a number of wood anatomical features setting them apart from *Bellucia* (Koek-Noorman et al., 1979). The similarities between the flowers of *Bellucia* and *Blakea* (and between those of other large-flowered melastome genera, see below) probably are the result of selection for a floral morphology optimally adapted to pollination by large bees (cf. POLLINATION AND DISPERSAL; Renner, 1986/1987).

Whereas *Bellucia* thus does not belong in the Blakeeae, it fits in well in all respects with the Miconieae, the only other neotropical berry-fruited tribe of the Melastomoideae; within the Miconieae it is particularly close to the genus *Loreya* DC. with which it has sometimes been united (cf. TAXONOMIC HISTORY and discussion below). *Myriaspora* also has flowers resembling those of *Bellucia*; however, in contrast to *Bellucia,* they have 5-lobed stigmas, 5-celled ovaries, 1-pored stamens, and petals that are strigulose outside. More significantly, *Myriaspora* differs from *Bellucia* in the smooth cuticle of the lower leaf surface and the more regularly tuberculate sculpture of the seed testa.

Bellucia, characterized by the (synapomorphic) papillose epidermis of the lower leaf surface, appears to be a well defined, monophyletic unit. Sterile material with mature leaves may be securely assigned to the genus because of the large size of the leaves, their leathery texture, shiny upper surface, and glaucous lower surface (due to the papillose epidermis). The related genus *Loreya* has leaves which are thinner, often brittle, and usually dark brown on the lower surface. Further characters distinguishing *Loreya* and *Bellucia* are: (1) *Loreya* is strictly 5-merous and has 5-lobed stigmas, with the exception of two species which occasionally have 6-merous flowers. *Bellucia,* on the other hand, is mostly 6- to 8-merous, except for one 5-merous species (the stigma of which, however, has 10 lobes). Also, flower mery is less fixed intraspecifically in *Bellucia.* (2) The bud petals in *Bellucia* (except in *B. huberi*) are completely covered until the last stages before anthesis, whereas in *Loreya* the bud petals outgrow the calyx which has only vestigial lobes. The petals are thus exposed at an earlier stage and for a much longer period. Also, *Bellucia* species have a broader tolerance of different edaphic and climatic conditions, they start to reproduce at an earlier age and flower and fruit almost continuously; in all these respects, *Loreya* is much more restricted. The two genera seem to have wood of different anatomy, specific weight, and color (however, the wood of only a few species has been investigated): that of *Bellucia* is relatively light, soft, and brittle, that of *Loreya* heavier and darker (Détienne & Jacquet, 1983; see VEGETATIVE ANATOMY). The low density of the wood in *Bellucia* may be correlated with the fast growth of these trees (cf. Liebermann et al., 1985). The combination of all these traits makes *Bellucia* weedier than *Loreya.* As pointed out in the section on DISTRIBUTION AND ECOLOGY, the species of *Bellucia* are pioneers adapted for colonizing newly opened habitats. Probably, *Bellucia* arose as an offshoot from *Loreya.* (See revision of *Loreya,* section on INTER- AND INTRAGENERIC RELATIONSHIPS, for a discussion of the possible paraphylesis of *Loreya.*)

The intent of the following is to incorporate the morphological, ecological, and geographical facts into a coherent picture of the evolution of *Bellucia.* The inferred order of divergence of the species of *Bellucia* is illustrated in Figure 4. Autapomorphies and parallelisms are discussed here, but not shown in Figure 4 since they provide no or little information on kinship. Assuming that *Bellucia* arose from an ancestral lineage morphologically similar to certain species of *Loreya,* the habitat of its ancestors would have been tropical lowlands perhaps in the southwestern subandean regions where *Bellucia* and *Loreya* are most diverse (compare Fig. 3 [*Bellucia*] and Fig. 15 [*Loreya*]). A spread into mid-elevation forests might have been correlated with the evolution of a more complete bud cover; today, *B. aequiloba,* with the thickest calyx lobes and strongly papillose and hairy lower leaf surfaces, occurs mostly at higher altitudes. These thick sepal lobes present a considerable investment in material; they may provide more protection against water loss through the petals than was afforded the exposed bud petals of a hypothetical ancestor with a truncate calyx like that of *Loreya* and many other genera in the Miconieae. The other evolutionary novelty of *Bellucia,* the papillose cuticle of the leaf lower surface where the stomata are, might also have originated as an adaptation for reducing transpiration. Protection of the flower buds may have been even more important during drier and cooler climatic conditions in the past; however, this argument may not apply in the case of *Bellucia* since the southwestern subandean region appears to have been among the climatically most stable areas of Amazonia.

If the present hypothesis is correct, the proto-*Bellucia* was a mid-elevation plant, which evolved a particular combination of characters, among them the complete bud cover and the papillose cuticle of the lower leaf surface. Subsequently, one group invaded the Amazonian lowlands where the stress factors of temperature and water deficiency were less important. In this lineage it became possible to reduce bud protection to a thin membrane and rely on growth-pressure (i.e., indirectly on water) to burst that membrane. At present, the species possessing a calyptra occur mainly in the lowlands. The calyptrate bud cover

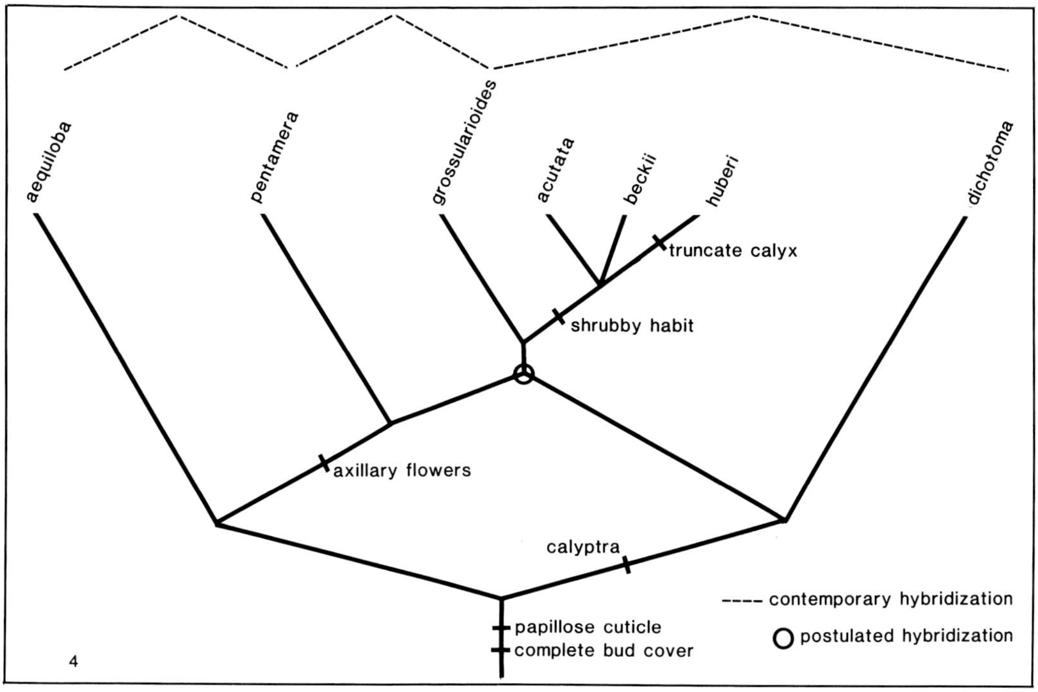

FIG. 4. Proposed phyletic relationships in *Bellucia.* See text for discussion.

involves less investment in terms of material and little complexity in terms of tissue differentiation (there are no lines of dehiscence). In the immediate vicinity of *Bellucia,* the calyptra is unique and clearly derived; within the Melastomataceae and Myrtales, calyptrate calyces have arisen several times.

One can envision further advantages to the calyptrate condition. For the breakage of the calyptra to occur, the water pressure in the petals must exceed a threshold. I have observed that during unusually dry spells, buds which normally would have opened the next day, halted development. Sometimes, such buds remained for two days with the calyptra slightly ruptured, but still protecting the petals from water loss. Anthesis starts only when a certain minimum of water is available to the plant, and therefore, the calyptra might be interpreted as a hydrostatic device, allowing for the fine-tuning of flowering times to water balance.

Of the two species which retain thick calyx lobes, one, *Bellucia aequiloba,* has branched inflorescences on the trunk, whereas the second, *B. pentamera,* has congested inflorescences which sometimes develop in the axils of existing leaves, sometimes on the older branches. Several related genera (*Henriettea, Henriettella, Loreya,* but not *Myriaspora*) have flowers on the old wood. Since several years must necessarily pass before the development of inflorescences on older wood, the axillary position of the flowers, which permits reproduction at an earlier age, should be of great selective value for pioneers; therefore, I consider it more advanced (within *Bellucia*!) than the cauline position. The 5-merous flowers and the strictly maintained floral mery of *Bellucia aequiloba* may be ancestral states still expressed in this species (the stigma, however, is 10-lobed). *Bellucia aequiloba* occurs mostly in primary forest and only rarely in advanced secondary vegetation. Figure 3 shows the restricted, perhaps relictual, range of the species.

Among the autapomorphies characterizing *Bellucia aequiloba* and *B. pentamera* are the scaly or flaky gray-brown bark of the branchlets in the former and the pronounced nodes and rows of pustule-like lenticels on the branchlets in the latter. Branched cauline inflorescences exactly like those of *Bellucia aequiloba* are also found in *Bellucia dichotoma.* However, this species has a calyptrate calyx and therefore belongs to the other lineage of the genus.

Bellucia grossularioides has axillary, 8-merous flowers with 13–14(–15) ovary cells, the highest number found in the genus. Evolutionarily, within *Bellucia,* the tendency probably was to increase flower mery since larger flowers permit a higher output of seeds, which likely correlates with increased weediness; indeed, *B. grossularioides* is extremely wide-spread (DISTRIBUTION AND ECOLOGY).

It seems likely that the shrubby species, *Bellucia acutata* and *B. beckii,* were derived from *B. grossularioides*-like ancestral stock because the three species share axillary flowers and calyptrate calyces. Autapomorphies of the shrubs are the lanceolate leaves in *Bellucia acutata* (all other species have broadly ovate leaves) and the rugose upper leaf surface in *B. beckii.* The third shrubby species, *Bellucia huberi,* has a truncate calyx and often axillary, but sometimes cauline flowers. If one assumes that *B. huberi* is closely related to the calyptrate, axillary-flowered species of *Bellucia,* the truncate calyx should be the result of progressive reduction of a former calyptrate calyx. The complete absence of even vestigial calyx lobes in *B. huberi* may indicate just this. Postulation of such a complete reduction of a former calyptra entails not having to assume the independent origin in *B. huberi* of the papillose lower leaf surface and seed morphology typical of *Bellucia.*

Whereas the presence of a truncate calyx in *Bellucia huberi* thus presents a problem (solved by according greater weight to the lower leaf surface and seed characters), the degree of pubescence in various other species provides a clear instance of parallel evolution. *Bellucia aequiloba, B. beckii,* and *B. dichotoma* share pilose to sublanate lower leaf surfaces and hypanthia (with simple hairs of one kind). The first species is a tree with thick, triangular calyx lobes and cauline inflorescences, the second is a shrub with axillary flowers and a calyptrate calyx, the third is a tree with cauline inflorescences and a calyptrate calyx; it is unlikely that their pilosity stems from an immediate common ancestor.

In only one of the arborescent species, *Bellucia pentamera,* is the axillary position of the flowers, an important adaptive trait (v.s.), not completely fixed; it may therefore have originated in the forerunners of this species. *Bellucia pentamera* does not have the calyptrate bud cover, but has the more costly, thick, triangular calyx lobes. The calyptrate line (Fig. 4), on the other hand, does have axillary flowers in one of its two arborescent members. The trait "axillary flowers" could have evolved twice, independently, in the two lineages. However, in view of the contemporary hybridization between calyptrate and non-calyptrate species (indicated by broken lines in Fig. 4; cf. CHROMOSOMES AND HYBRIDIZATION), an ancient hybridization event by which the once-evolved advantageous trait "axillary flowers" spread from one lineage to the other seems more likely than its parallel development. This is the reason why I am postulating a hybridization event in the phylogenetic tree (Fig. 4) between the lineage retaining thick calyx lobes and that which evolved the presumably advantageous calyptrate condition. The pattern of character states in *Bellucia* may thus be an outcome of reticulate evolution.

TAXONOMIC TREATMENT

Generic Description

Bellucia Rafinesque, Sylva tellur. 92. 1838, nom. cons. Type: *Bellucia grossularioides* (L.) Triana.

Webera J. F. Gmelin, Syst. nat. **2(1):** 776, 820. Sep–Nov 1791, *non* Schreber, May 1791, nec Hedwig, 1801. Type: *Webera quinquenervia* (Aublet) J. F. Gmelin (*Blakea quinquenervia* Aublet).
Apatitia Desvaux *in* Hamilton, Prodr. pl. Ind. occid. 42. 1825. Type: *Apatitia blakeoides* Desvaux.
Axinanthera Karsten, Linnaea **30:** 157. Post Apr 1859. Type: *Axinanthera macrophylla* Karsten.

Small or medium-sized *trees,* to 25 (rarely 30) m high or *shrubs*; trunk cylindrical; bark scaly, flaky, with longitudinal fissures or smooth. *Leaves* opposite, simple, petiolate, large, coriaceous and glossy green when fresh, drying brown above, glaucous below, 3–5-pli-nerved, the inner pair of lateral primary veins departing 0.5–3 cm above the blade base, margin entire or serrulate (when young), lower leaf surface pilose, villose, or sublanate with smooth, thin, sometimes slightly curled trichomes or both surfaces glabrous. *Inflorescences* 2–3-branched cymes or congested clusters of few flowers, rarely one, in the axils of existing leaves, on the branches below the existing leaves, or on the trunk. *Flowers* hermaphroditic, 5–8-merous; hypanthium subglobose; calyx a calyptra, splitting open in an irregular semicircle and drying as a hyaline membrane, often persistent on the limb after anthesis, or calyx dehiscing regularly into triangular lobes, lobes valvate in bud, erect and persistent after anthesis, or calyx limb truncate (in *B. huberi*); petals obovate with a claw, apically truncate or slightly emarginate, fleshy, adaxially with two longitudinal and one transversal callus ridge, spreading at anthesis, completely white or white inside and pink-flushed outside; stamens twice as many as the petals, isomorphic, glabrous, laterally compressed, anther and filament of about equal length; anthers thick, with two slightly introrse pores, the connective not prolonged but fleshy, dorsi-basally yellow; ovary completely inferior and fused with the hypanthium wall, 10–14(–15)-locular; style glabrous, thick, terete, straight or slightly sigmoid, longer than the stamens; stigma capitate and 10–16-lobed. *Fruit* a many-seeded, semi-globose, glabrous or rarely hirsute, greenish-yellow berry. *Seeds* minute, 0.5–1 mm long, more or less ovoid, with a small, subterminal hilum; testa tuberculate.

A neotropical genus of seven species, ranging from southern Mexico through Central America to northern Bolivia and east through the Guianas to Maranhão and Bahía in Brazil; lowlands to 1600 m.

Key to the Species of *Bellucia*

1. Calyx dehiscing regularly into persistent triangular lobes or calyx truncate.
    2. Calyx truncate, ca. 2 mm long, buds completely smooth; petals ca. 12 mm long. 7. *B. huberi.*
    2. Calyx dehiscing regularly into persistent triangular lobes, 5–8 mm long, lines of dehiscence visible in buds; petals 15–18(–20) mm long.
        3. Lower leaf surface with reddish villose to sublanate pubescence; inflorescences 2–3-branched, many-flowered, borne on the branches and the trunk. 1. *B. aequiloba.*
        3. Lower leaf surface glabrous; inflorescences congested, few or 1-flowered, axillary or sometimes on the trunk. 2. *B. pentamera.*
1. Calyx a calyptra dehiscing in an irregular semicircle, drying as a hyaline membrane, usually persistent on the limb, buds apically completely smooth.
    4. Trees; leaves ovate to broadly ovate 15–35 × 10–20 cm; petals 20–24 mm long.
        5. Cymes 2–3-branched, borne on branches below the leaves or on the trunk; lower leaf surface reddish villose. 3. *B. dichotoma.*
        5. Cymes usually 2-flowered, in the axils of existing leaves; lower leaf surface glabrous. 4. *B. grossularioides.*
    4. Shrubs; leaves oblong to elliptic, 11–18 × 4–11.5 cm; petals 12–15 mm long.
        6. Leaves lanceolate (oblong), lower surface glabrous. 5. *B. acutata.*
        6. Leaves elliptic, lower surface with sublanate pubescence. 6. *B. beckii.*

1. Bellucia aequiloba Pilger, Verh. Bot. Vereins Prov. Brandenburg **47:** 182. 1905. Type: Brazil. Acre: Juruá-Mirím, Cruzeiro do Sul, Aug 1901, *Ule 5664* (lectotype here designated: HBG!; isolectotypes: G!, MG!, F photo neg. 17249!). Holotype at B destroyed, but represented by a photo, F neg. 17249!

Figs. 1E, 5A–D.

Tree, (6–)10–21 × 0.10–0.25 m; bark scaly, flaky, gray-brown. *Leaves* 21–42 × 12–39 cm, broadly ovate, shortly acuminate, rounded at base, blades with the inner pair of lateral primaries departing ca. 1 cm above the base, lower leaf surface reddish villose to sublanate; petioles 3–6.5 cm long. *Cymes* 2–3-branched, on branches below the leaves and on the trunk, reddish pilose; pedicels 5–12 mm long. *Flowers* 5-merous; hypanthium ca. 9 mm long, reddish pilose; calyx dehiscing into five thick, triangular lobes, 6–8 mm long, persistent; petals 15–18 (–20) mm long, white, pink-flushed outside; filaments 8–9 mm long, thecae 6–7 mm long; ovary 10-locular; style 20–22 mm long. *Fruit* a greenish-yellow berry. *Seeds* ca. 0.7 mm long, ovoid, irregularly tuberculate and with shallow grooves.

Distribution and ecology (Fig. 6): *Bellucia aequiloba* occurs in the southern subandean part of the Amazon basin in seasonally inundated forest along rivers; in La Paz province in Bolivia, it inhabits submontane forest at elevations of 300–1500 m; it is occasional in advanced secondary vegetation.

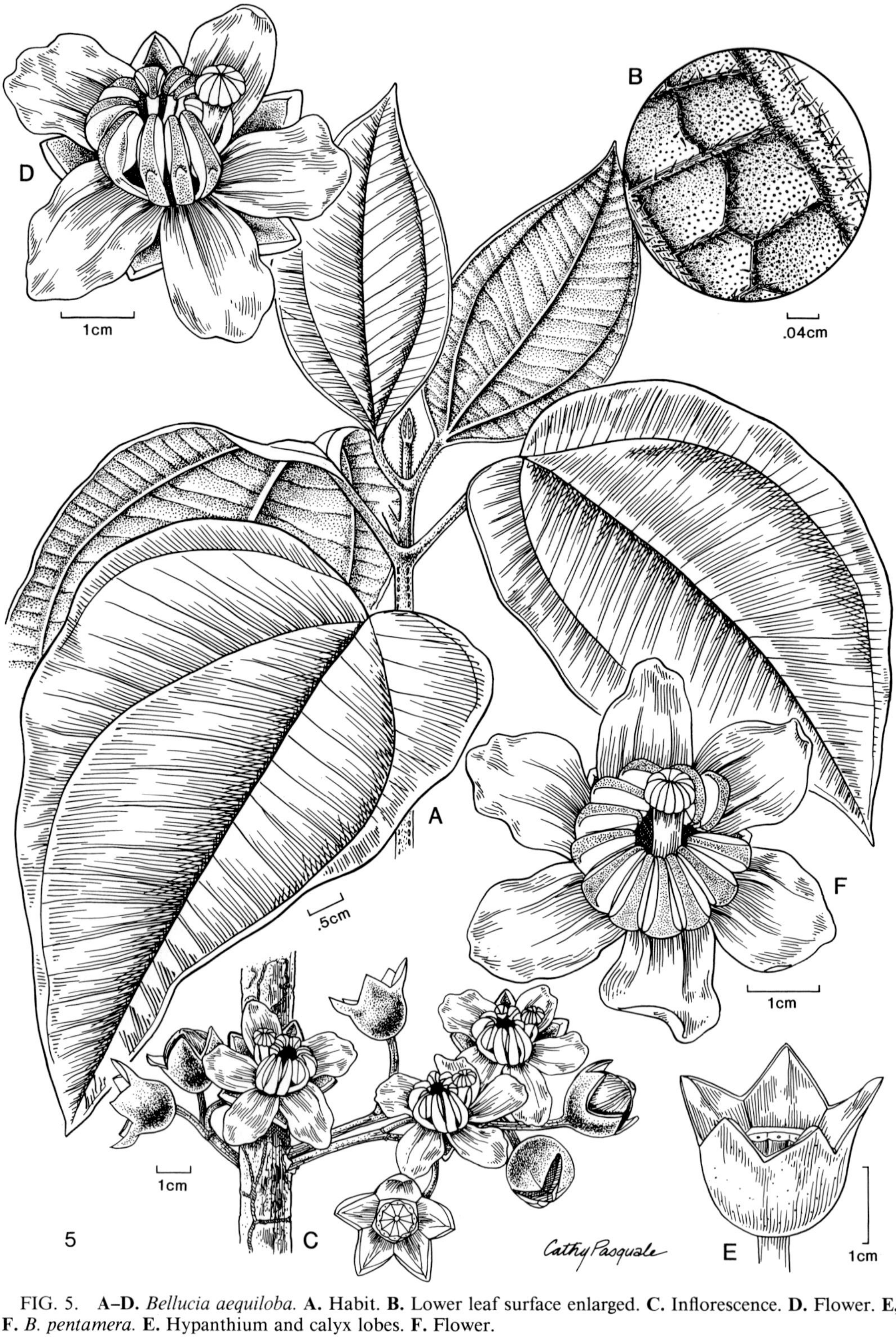

FIG. 5. **A–D.** *Bellucia aequiloba.* **A.** Habit. **B.** Lower leaf surface enlarged. **C.** Inflorescence. **D.** Flower. **E, F.** *B. pentamera.* **E.** Hypanthium and calyx lobes. **F.** Flower.

Representative specimens examined. PERU. HUANUCO: Pachitéa, Dtto. Puerto Inca, *Schunke 2908* (COL, F, G, NY, VEN). MADRE DE DIOS: Tambopata, Dtto. Tambopata, Maldonado–Quincemil rd., km 12.5, *Malpica 47* (MO).

BRAZIL. ACRE: Rio Branco–Porto Velho rd., km 22–42, *Albuquerque 1277* (MO, NY, US); 15 km S of Cruzeiro do Sul, *Cid et al. 5196* (INPA); Igarapé perto Tarauacá, *Miranda 7* (INPA); Porangaba, Rio Juruá-Mirím, *Prance et al. 12980* (COL, F, INPA, MG, NY, RB, S, SP, US). AMAZONAS: Track from Boca do Acre airstrip to Monte Verde, north bank of Rio Purús, *Prance et al. 2483* (F, INPA, MG, NY, R, S, U, US, W).

BOLIVIA. BENI: Prov. Vaca Diez, vic. of Chácobo village, 11°45′S, 66°02′W, *Boom 4490* (LBP, NY, U, US). LA PAZ: Between Tipuani and Guanai, *Bang 1645* (BM, BR, F, G, G-BOISS, M, MO, NY, US); Prov. Laracaja, Consata, 7 km toward Mapiri, *Beck 4904* (LPB, US); nr. Mapiri, San Carlos, *Buchtien 1033* (NY); vic. cataracts of the Boopi river, *Rusby 722* (NY); Tumupasa, *R. S. Williams 576* (BM, NY, US).

Local names: nispero (Peru), jambo (Brazil), peras, huicama (Chácobo indians, Bolivia).

A collection from Pando province in Bolivia (*Fernández Casas & Susanna 8201*) suggests hybridization between *Bellucia aequiloba* and *B. pentamera*; the specimen has much-branched inflorescences borne on the old wood like those of typical *B. aequiloba,* but in the relatively short length of its sepals and the scarce pubescence of the lower leaf surface it resembles *B. pentamera.*

2. **Bellucia pentamera** Naudin, Ann. Sci. Nat. Bot. III, **16:** 105. 1850c. Type: Peru. Cuzco: Nr. town of Cuzco, *Gay 1269* (holotype: P!, F photo neg. 36338!; isotype: P!).
Figs. 1C; 2J; 5E, F.

Axinanthera macrophylla Karsten, Linnaea **30:** 157. Post Apr 1859; *Bellucia axinanthera* Triana, Trans. Linn. Soc. London **28:** 142. 1871, nom. illegit. (according to Art. 63 of I.C.B.N.). Type: Colombia. Meta: Nr. Villavicencio, *Karsten s.n.* (holotype: LE; isotypes: F!, W!).

Bellucia costaricensis Cogniaux, Bull. Soc. Roy. Bot. Belgique **30(1):** 264. 1891b. Type: Costa Rica. Alajuela: Nr. Buenos Aires, *Pittier 3916* (lectotype here designated: BR!; isolectotypes: BR!, US!).

Bellucia weberbaueri Cogniaux, Bot. Jahrb. Syst. **42:** 148. 1908. Type: Peru. San Martín: Moyobamba, Aug 1904, *Weberbauer 4505* (holotype: BR frag.!). Isotype at B destroyed, but represented by a photo, F neg. 17250.

Bellucia aricuaizensium Pittier, Explor. Cuenca de Maracaibo 40. 1923. Type: Venezuela. Zulia: Río Lora, 14 Dec 1922, *Pittier 10956* (holotype: US!; isotype: NY!).

Shrub or *tree,* 3–15 × 0.10–0.15 m; bark brown, fissured, on the twigs scaly; twigs with pronounced nodes and longitudinal rows of pustule-like lenticels. *Leaves* 20–35(–40) × 12–28 cm, broadly ovate, shortly acuminate, broadly acute at base, with the inner pair of lateral primaries departing 1–2 cm above the base, glabrous; petioles 2–5 cm long. *Cymes* congested, few or 1-flowered, on the trunk or branches below the leaves or in the axils of existing leaves, glabrous; pedicels 10–15 mm long. *Flowers* 5–6(–7)-merous; hypanthium ca. 10 mm long, glabrous; calyx dehiscing into (4–)5–6 triangular lobes, 5–7 mm long, persistent; petals 15–18(–20) mm long, white, sometimes pink-flushed outside; stamens 10(–11–)12(–13–)14, filaments 5–7 mm long, thecae ca. 6 mm long; ovary 12–13(–14)-locular; style 20–22 mm long. *Fruit* a greenish-yellow, subglabrous berry. *Seeds* ca. 0.6 mm long, ovoid, irregularly tuberculate and with shallow grooves.

Distribution and ecology (Fig. 9): *Bellucia pentamera* occurs from southern Mexico south, along both sides of the Andes, to Bolivia and to the eastern periphery of the Amazon basin in Brazil. It usually grows on non-flooded ground, but sometimes in riverine forest and is twice reported from regularly inundated (várzea) vegetation; lowlands to 1650 m elevation.

Representative specimens examined. MEXICO. TABASCO: Mun. Huimanguillo, km 32 de la desviación de Huimanguillo hacia Fco. Rueda, *C. Cowan 3121* (CAS, NY). VERACRUZ: Mun. Hidalgotitlán, en El Ejido Agustín Melgaz, a 2 km del campto. Hnos. Cedillo, *Calzada 7358* (F); Mun. Jesús Carrzana, Juan de la Luz Enriquez, 17°27′N, 95°03′W, *Castillo 401* (F, US); Mun. Hidalgotitlán, lomitas al SE del poblado, *Wendt et al. 2836* (CAS, US).

GUATEMALA. ALTA VERAPAZ: Semicoch, *Hay s.n.* (US). IZABAL: Mariscos, on La Minera rd., 3 km SE, *Contreras 7591* (CAS, VEN); St. Thomas, *Deam 6050* (MO, US); 1–2 km S of Izabal, 15°15′–15°35′N, 84°00′–84°25′W, *Jones et al. 3018* (F, NY, US); Morales, *Kellerman 6108* (US); nr. Entre Ríos, *Kuylen 146* (US); at Shell Station and vic. just S of Río Dulce, *Le Doux et al. 2090* (NY); Las Quebradas, *Pittier 8572* (NY, US); S facing lower ridges of Cerro San Gil, opposite Cayuga, *Steyermark 39476* (F, NY).

BELIZE. TOLEDO DISTR.: Cohume ridge, nr. Crique Queva, *Gentle 5587* (F, S, US).

HONDURAS. COLON: Cuyamel, *Carleton 579* (US). GRACIAS A DIOS: La Mosquita, alrededores del Río Plátano, 15°30′–55°N, 84°00′W, *Clewell & Cruz 4123* (MO, US); Las Marias, Renare Station, camp Tiro, 2 mi NW of Balebar on 3rd northern branch of Quebrada Tiro, tributary of Río Plátano, 15°43′N, 84°50′W, *Saunders 1075B* (BM, F, NY).

NICARAGUA. ZELAYA: Between El Muelle de los Bueyes and Villa Somoza, town nr. limit of Dept. Chontales, *Bunting & Licht 1041* (F, NY, US); Awas Tingni, 14°23′N, 83°57′W, *Little 25141* (MO, US).

COSTA RICA. ALAJUELA: Savanna nr. Buenos Aires, *Pittier 4955 = Tonduz 4955* (BM, BR, F, G, P, US). PUNTARENAS: Rd. to San Vito, between Paseo Real and San Vito, *Lumer 1363* (NY). SAN JOSE: Edge of the Río Peñas Blancas and area around the property of A. Skutch, 9°20′N, 83°34′W, *Burger & Matta 4825* (F, NY, US, W).

PANAMA. BOCAS DEL TORO: Río San Pedro, *Gordon 30C* (MO). CANAL ZONE: Pipeline rd., *Sullivan 88* (CAS). CHIRIQUI: Vic. Remedios, *Allen 3669* (BR, G, MO, NY, P). COCLÉ: Along Río San Juan below its jct. with Río Tife, *Hammel 3402* (CAS, MO). COLON: María Chiquita, *Dwyer 5195* (MO). DARIEN: Vic. of Cana, *Stern et al. 672* (G, MO, NY, US). PANAMA: Rd. S-10 north of Escobal nr. jct. with rd. S-1. *Croat 12485* (F, MO, NY).

GUADELOUPE. End of rd. to Sofaya, above St. Rose, N end of island, *Holdridge 440* (NY).

COLOMBIA. AMAZONAS: Quebrada Arara, 2 hr N of Leticia nr. Amazon river, *Croat 7534* (MO). ANTIOQUIA: Río Samana, *Daniel 5537* (COL, US); Mpio. San Luis, cerca de una quebrada, *Loaiza et al. 174* (HUA, JAUM, VEN). CAQUETA: 13 km SE of Morelia, along rd. to Río Pescado, SW of Florencia, *Davidse 5674A* (COL, MO). CAUCA: Sta. Bárbara, Timbiquí, *Lehmann 559* (NY); Río Calima, márgen izquierda, bosques junto a Quebrada de la Brea, *Cuatrecasas 21136* (COL, F, NY, US). CHOCO: 6–10 km de Tutunendo, carretera Medellin-Quibdo, *Gentry & Fallen 17532* (AAU, COL, MO, NY). CUNDINAMARCA: La Palma, carretera a Pacho, Río Murca, *García-Barriga 12400* (COL, US). META: Villavicencio, *Cuatrecasas 3522* (COL, F, US); Sierra de la Macarena, Caño Yerly, *Philipson et al. 1550* (BM, COL, US). NARIÑO: Barbacoas, corregimiento Santander (Buenavista) a Barbacoas (vertiente del Río Telembí), *García-Barriga 13142* (COL, US). NORTE DE SANTANDER: La Motilonia, frontera Colombo–Venezolana, hoya del Río de Oro, 9°10′N, 72°50′W, *García-Barriga & Lozano 18275* (COL, NY, P). PUTUMAYO: Uplands of Río Putumayo, about Quebrada Negra, vic. of Puerto Asis, ca. 0°30′N, 76°30′W, *Ewan 16753* (BM, NY, US). SANTANDER: Río Carare, San Juan, *Rentería et al. 1489* (COL, MO, NY). TOLIMA: Entre Honda y Samana, Río La Miel, *Pérez-Arbeláez 10336* (COL, NY).

VENEZUELA. AMAZONAS: Casiquiare, Vasiva et Pacimoni, *Spruce 3212* (BR, G, G-BOISS, G-DC, GOET, K, MG, NY, P, W), *Spruce 3268* (BR, G, G-BOISS, G-DC, GOET, IAN, NY, P, W); along river just below mouth of Río Cunucunuma, *Wurdack & Adderley 43080* (NY, US, VEN). APURE: Reserva Florestal San Camillo, a lo largo del Río Nulita, N del caserío San Camilo (El Nula), *Steyermark et al. 101316* (P, US, VEN). BARINAS: Along Cid. Bolívar–Sta. Bárbara de Bariñas rd., *Aristeguieta 1586* (VEN). BOLIVAR: Distr. Sucre, entre el caserío Las Trincheras y el cementerio, *Ruiz-Terán et al. 10695* (US). CARABOBO: Las Vegas, El Trompillo, May 1970, *Pantin s.n.* (US, VEN). COJEDES: Manrique, *Delascio 4394* (VEN). DELTA AMACURO: Between Amacuro and mouth of Deadwater Creek Moat, *Steyermark 87386* (F, M, NY, U, US, VEN). MERIDA: Along Río Caparo, ca. 1 km from dam site, ca. 07°41′N, 71°28′W, *Liesner & González 9370* (MO, VEN). MONAGAS: Reserva Florestal de Guarapiche, Cañon Colorado, *Aristeguieta et al. 7194* (US, VEN). TACHIRA: Ca. 2 km SE of San Juan de Colón, *Bruijn 1334* (F, MO, NY, U). TRUJILLO: Dtto. Baralt, Maracaibo–Agua Viva rd., 46 km N de Agua Viva, 1 km S de La Raya, *Bunting 5019* (US, VEN). YARACUY: Sierra de Aroa, Cerro Tigre, 10 km E of Aroa air distance, Río Carabobo and adjunct slopes up to rd., 10°26′N, 68°49′W, *Liesner & González 9702A* (MO, US, VEN). ZULIA: Distr. Perijá, ca. 7 km E of the Maracaibo–La Fria Hwy. (Hwy. 6) and ca. 4 km N of the Río Aricuaisa, 9°26′N, 72°33′W, *Davidse et al. 18331* (MO, NY, VEN).

ECUADOR. CARCHI: Around Maldonado, *Madison et al. 4801* (F, QCA, SEL, US). ESMERALDAS: Imbabura, Lita, at the boundary to Esmeraldas and Carchi, ravine of Río Lita, *Harling 4303* (NY, S). MORONA–SANTIAGO: SE from Misión Bomboiza, across rd. and stream, *MacBryde 157* (MO, QCA, US). NAPO: Veracrúz, *Asplund 19446* (B, NY). PASTAZA: Canelos, *Lugo 1585* (MO, NY, S, US). ZAMORA-CHINCHIPE: Zamora–Gualaquiza rd., ca. 5 km N of Cumbaraza, *Harling & Andersson 13767* (GB, MO, US).

PERU. AYACUCHO: Puerto San Francisco, Prov. La Mar, *Saavedra S. 6337* (US). AMAZONAS: Pongo de Manseriche, Marañon, ca. 77°30′W, *Tessmann 3523* (G, NY). HUANUCO: Prov. Pachitéa, Dtto. Honoria, a lo largo del Río Pachitéa, cerca del campamento Miel de Abeja, 1 km arriba de la confl. con el Río Ucayáli, *Schunke 2237* (F, G, NY, US). JUNIN: San Francisco de Satipo, hillsides along Río Satipo, *Solomon 3238* (MO). LORETO: Prov. Maynas, Dtto. Alto Nanay, nr. Sta. María de Nanay, *Simpson & Schunke 708* (F, G, NY, US). SAN MARTIN: Alone rd. between Moyobamba and Chachapoyas, vic. of km markers 430–31, just E of village of Naranjos at Río Naranjos, 5°21′S, 77°20′W, *Croat 58160* (MO, US).

BRAZIL. ACRE: Nr. mouth of Rio Macauhan, tributary of Rio Iaco, 9°20′S, 69°W, *Krukoff 5580* (BM, F, G, M, MO, NY, RB, U, US); Serra da Moa, *Prance et al. 12666* (F, INPA, MG, NY, P, RB, SP, US); Rio Branco, behind Parque Zoobotânico, *Renner 990* (INPA, SP). AMAPÁ: Cachaco, Serra do Navío, *Rabelo 333* (MG, NY). AMAZONAS: Humaitá, *Renner 502* (HBG, INPA, US); São Paulo de Olivença, nr. Palmares, *Krukoff 8057* (B, BM, BR, F, G, MO, NY, U). GOIÁS: 10 km S of Guarai, *Irwin et al. 21497* (F, MO, NY, UB, US). MARANHÃO: Rio Gurupí, *Fróes 34233* (IAN). MATO GROSSO: Xavantina, estr. para São Felix, *Fonseca 363* (NY, US). PARÁ: BR 22 (Capanema–Maranhão) km 70, *Prance & Silva 58857* (F, MG, NY, U, UB, US). RONDÔNIA: Source of Jatuarana river, Machado river region, *Krukoff 1572* (BM, G, MO, NY, P). RORAIMA: Canto Galo, Rio Mucajaí between Pratinha and Rio Apiaú, *Prance et al. 3975* (F, INPA, MG, NY, R, RB, U, US).

BOLIVIA. BENI: Along Río Yata, ca. 40 km SW of Guajará-Mirím, *Anderson 11968* (US). LA PAZ: San Carlos nr. Mapiri, *Buchtien 1738* (NY).

Local names: pomarrosa, galleta (Mexico), túndkia, manzana de montaña (Guatemala), coronillo (Costa Rica), oreja de mula, uráno (Guaymi indians, Panama), coronillo, nispero, guayavo de pava (or paba or fara), guayavo-pomo (Colombia), túnguia (Ecuador), túnkia, túgkia, bispera, dispero, nispero, yujash, yuháč, yuhís (Peru), araçá de anta, goiaba de anta, jambo do mato (Brazil).

Bellucia pentamera was introduced successfully on Guadeloupe and at Bogor in Java (Helten, 1923; van Steenis, 1975), and is now spontaneously regenerating in Java, Sumatra, Borneo, and West Malaysia (Stone, 1972). Prior to 1953 it was brought to the Congo area where it became widely naturalized. Apparently, an attempt to introduce this species in Surinam failed: repeated collections from a single plant (tree *no. 1654, B.W. herb. nos. 1985, 2508, 2655, 2725, 5005, 5571*) growing in Watramiri south of Paramaibo during the years 1916–1920 remain the only record of the species from the Guianas.

A number of collections appear to be hybrids between *Bellucia pentamera* and *B. grossularioides.* The specimens possess flowers on the branches below the leaves as in *B. pentamera.* Their calyces, however, differ from that of typical *B. pentamera* in that the lobes do not split all the way and are apically hyaline, thus resembling the calyptra of *B. grossularioides.* In some flowers, three or four lobes may be completely connate, whilst a fifth may be a normal triangular sepal. In their bark, these specimens resemble *B. pentamera,* i.e., the bark is light tan colored and scaly or flaky, but in number of stamens they are intermediate between the parents having mostly 14 stamens; pure *B. pentamera* has 10–*12*–14 and pure *B. grossularioides* (14–)16–18. *Oliveira 1400,* from Goiás, Brazil, is almost certainly such a hybrid. The collections *Steyermark & Luteyn 129711* from Amazonas, Venezuela (from the same tree as *Thomas 3375* and *Renner s.n.*; flowers in FAA in my collection), *Gentry et al. 18585* and *Díaz 954* from Loreto, Peru, and *Sastre 2008* from Guadeloupe, also appear to come from hybrids. In each case, both parents are known to occur in the area where the hybrids have been collected. A putative hybrid between *B. pentamera* and *B. aequiloba* from Bolivia (*Fernández Casas & Susanna 8201*) has already been mentioned under the latter species.

Bellucia pentamera is illustrated in the *Flora Columbiae* (Karsten, 1861), Mutis' *Flora de la Real Expedición* (1760–1817; 1983), Stone (1972), and in the floras of Panama (Gleason, 1958), Guatemala (Standley & Williams, 1963), Cundinamarca (Uribe Uribe, 1972), and Ecuador (Wurdack, 1980a).

3. Bellucia dichotoma Cogniaux *in* Martius, Fl. bras. **14(4):** 514. 1888. Type: Brazil. Pará: Santarém, Apr 1850, *Spruce 819* (lectotype here designated: P!, photo F neg. 36841!; isolectotypes: G-BOISS!, G-DC!, K!, photo, H. A. Gleason s.n.!, P!, W!). Figs. 1D, 2G.

Bellucia imperialis Saldanha & Cogniaux *in* Martius, Fl. bras. **14(4):** 515, pl. 109. 1888. Type: Brazil. Amazonas: Manaus, *Schwacke III.n.374* (=*Glaziou 13830*) (lectotype here designated: BR frag.!; isolectotypes: GOET!, P!, R!).

Tree, 4–8(–12) × 0.07–0.20 m; bark brown, with fine cracks and pustules. *Leaves* 21–30 × 14–20 cm, broadly ovate, shortly acuminate, subrotundate at base, the inner pair of lateral primaries departing ca. 1 cm above the base, lower leaf surface reddish villose, pilosity quite variable in density; petioles 2–5 cm long. *Cymes* 2–3-branched, on the trunk and on the branches below the leaves; pedicels 10–15 mm long. *Flowers* mostly 6-merous; hypanthium 10–12 mm long, pilose; calyx a calyptra, dehiscing in an irregular semicircle, drying as a hyaline membrane, often persistent on the limb after anthesis; petals 20–22 mm long, white; stamens (11–)12 (–14), filaments 6–8 mm long, thecae ca. 5 mm long; ovary (10–)11–12-locular; style ca. 20 mm long. *Fruit* a greenish-yellow, subglabrous berry. *Seeds* ovoid, ca. 0.5 mm long, testa irregularly tuberculate.

Distribution and ecology (Fig. 7): *Bellucia dichotoma* occurs in the central and lower Amazon basin in non-flooded forest; it grows in disturbed and undisturbed vegetation on clay loam or sandy soil.

Representative specimens examined. BRAZIL. AMAPÁ: Lower slopes and base of Mt. Tipac, Rio Oiapóque, 3°36′N, 51°19′W, *Irwin 48756* (F, IAN, MG, NY, U, UB, US). AMAZONAS: Manaus–Itacoatiára rd., km 26, Ducke reserve, *Prance et al. 2161* (COL, F, INPA, MG, NY, R, U, US); Manaus, INPA campus, *Renner 8* (HBG, INPA, US); Mun. Borba, mouth of Rio Canúma, across river from Canúma, 4°S, 59°W, *Todzia et al. 2264* (NY). PARÁ: Itaitúba, *Ducke 2952* (BR, MG); Belém, *Huber 123* (BR, F, G-BOISS, MG); Santarém, nr. airport, *Renner 983* (INPA, UB); Rio

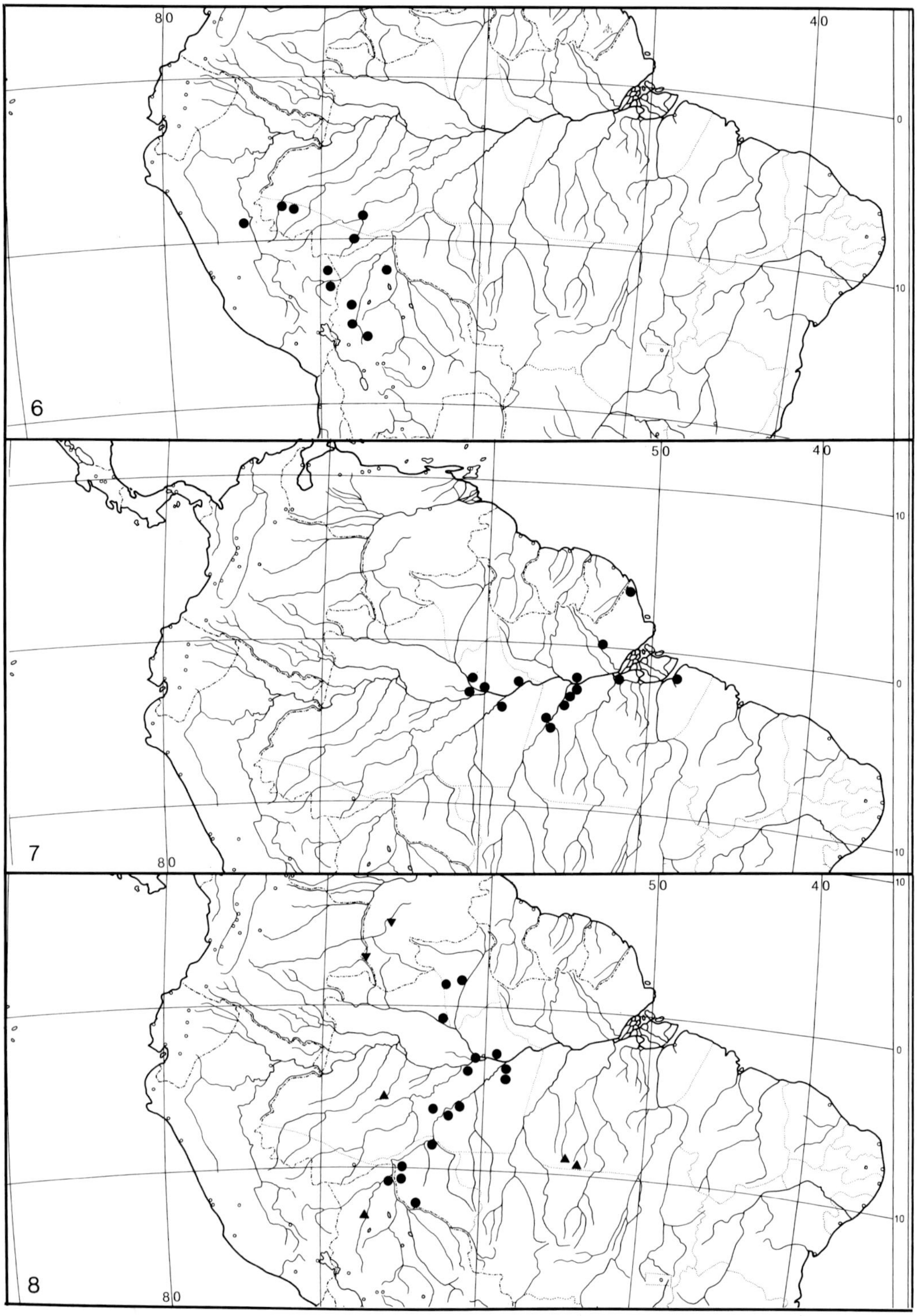

FIGS. 6–8. Distribution of *Bellucia*. **6.** *B. aequiloba*. **7.** *B. dichotoma*. **8.** *B. acutata* (●), *B. beckii* (▲), and *B. huberi* (▼).

Xingú, Porto de Moz, *Silva 1510* (MG); Tapajóz, Villa Braga, *Snethlage s.n. = MG no. 10054* (MG).

Local names: araçarana, araçá de anta, papa terra vermelha, papa terra amarela (Brazil).

The bark is used to waterproof ropes and wooden surfaces; it is also thought to have medical properties as an infusion.

The characters used by Cogniaux to separate *Bellucia imperialis* from *B. dichotoma*, both based on Spruce collections, do not hold. In fact, Spruce (1908, vol. **1:** 223) regarded the material he sent from Santarém and that from Manaus as conspecific. Artificial pollinations carried out in Manaus with pollen of a flower brought from Santarém the day before showed that the two populations were interfertile; big fruits with viable seeds resulted from the experiment. Interspecific hybrids between *B. dichotoma* and *B. grossularioides* are discussed under the latter species.

Bellucia dichotoma was once introduced in the Botanical Garden in Rio de Janeiro but does not persist. Several persons who have tried to grow the species as an ornamental in southern Brazil assert that it does not survive outside its central Amazonian habitat. Macbride (1941) misattributed *Bellucia dichotoma* (sub *B. imperialis*) to Peru by confusing it with *B. pentamera*.

Bellucia dichotoma is illustrated in Cogniaux (1888, pl. 109).

4. Bellucia grossularioides (Linnaeus) Triana, Trans. Linn. Soc. London **28:** 141. 1871. Figs. 1A, 2D.

Melastoma grossularioides Linnaeus, Sp. pl. 390. 1753, and Sp. pl. ed. 2. 558. 1762–1763, where Linnaeus removed the Plumier element. Lectotype here designated: Plukenet, Phytographia 3, pl. 249, fig. 4. 1692.

Blakea quinquenervia Aublet, Hist. pl. Guiane **1:** 525, pl. 210. 1775; *Bellucia quinquenervia* (Aublet) Karsten, Linnaea **30:** 159. Post Apr 1859; *Blakea triplinervia* L. f., Suppl. 246. 1782, nomen illeg.; *Webera quinquenervia* Gmelin, Syst. nat. **2(1):** 820. Late 1791; *Bellucia nervosa* Rafinesque, Sylva tellur. 93. 1838, nomen illeg. *Bellucia aubletii* Naudin, Ann. Sci. Nat. Bot. III, **16:** 102. 1850c, nomen illeg. Type: French Guiana. Nr. Aroura, *Aublet s.n.* (holotype: BM!; isotype: LINN-SM, fide Howard, 1983).

Blakea macrophylla D. Don, Mem. Wernerian Nat. Hist. Soc. **4:** 326. 1823; *Bellucia macrophylla* (D. Don) Triana, Trans. Linn. Soc. London **28:** 142. 1871. Type: Mexico. *Pavón s.n.* (holotype: BM).

Blakea macrophylla D. Don is based on a Sessé and Moçiño collection from Mexico, which was sold to Lambert whose assistant, David Don, described it as a new species. I have not seen the type specimen, but John J. Wurdack (pers. comm.) saw it at BM in 1975; at the time, he did not make notes on the structure of the calyx, which would permit placement of the specimen in either *B. grossularioides* (with a calyptrate calyx) or *B. pentamera* (with triangular sepals). The latter is the only other species of *Bellucia* occurring in Mexico. Of the sketches Sessé and Moçiño had drawn for the plates of their *Flora Mexicana* (ined.) from living specimens in the field (McVaugh, 1982), one clearly shows a 7-merous flower of *B. grossularioides* with the typical, irregularly split, calyptrate calyx. A slide of this sketch (sketch no. 1973, fig. 75.II in the Torner Collection) was kindly made available to me by the Hunt Institute for Botanical Documentation. The accompanying Spanish text states ". . . flores de dos en dos en las axilas, y en los ramos, las divisiones del calix blanquecinas." This refers to the typically geminate axillary flowers of *B. grossularioides* and probably to the irregularly splitting calyptra, the remnant of which often persists as a shrivelled pale or hyaline membrane.

Apatitia blakeoides Desvaux *in* Hamil., Prodr. pl. Ind. occid. 42. 1825. Type: French Guiana. *Perrottet s.n.* (lectotype here designated: P!).

Bellucia hostmannii Naudin, Ann. Sci. Nat. Bot. III, **16:** 104. 1850c. Type: Surinam. *Hostmann 756a* (holotype: P!; isotypes: C!, G!, G-BOISS!, GOET!, MO!, NY frag.!, U!, W!).

Bellucia superba Naudin, Ann. Sci. Nat. Bot. III, **16:** 104. 1850c. Type: Mexico. Tabasco: El Azufre nr. Teapa, *Linden s.n.* (holotype: P!; isotype: G!).

Bellucia brasiliensis Naudin, Ann. Sci. Nat. Bot. III, **16:** 104. 1850c. Type: Brazil. Amazonas: *Ferreira s.n.* (holotype: P!; isotype: P!).

Bellucia multiflora Karsten, Linnaea **30:** 158. Post Apr 1859. Type: Colombia. Meta: Nr. Villavicencio, *Karsten s.n.* (holotype: LE: isotypes: F!, NY!, P!, W!).

Bellucia circumscissa Spruce ex Cogniaux *in* Martius, Fl. bras. **14(4):** 515, pl. 108. 1888. Type: Brazil. Amazonas: São Gabriel, *Spruce 2162* (holotype: BR frag.!; isotypes: BM!, C!, NY photo neg. 21161!, G!, G-BOISS!, GOET!, M!, NY!, P!, US!, W!).

Tree, 3–8–25(–30) × 0.10–0.30(–1) m; bark gray or brown, smooth with fine, shallow, vertical fissures. *Leaves* 15–35 × 10–20 cm, broadly ovate, shortly acuminate, obtuse to rounded at base, with the inner pair of lateral primaries departing ca. 1 cm above the base, glabrous; petioles to 6 cm long. *Cymes* congested, usually 2-flowered, in the axils of existing leaves, glabrous; pedicels 10–24 mm long. *Flowers* (6–)7–8(–9)-merous; hypanthium 8–10 mm long, glabrous; calyx a calyptra, dehiscing in an irregular semicircle, drying as a hyaline membrane, often persistent on the limb after anthesis; petals (20–) 22–24 mm long, white, sometimes pink-flushed outside; filaments 7–9 mm long, thecae 5–6 mm long; ovary (12–)13–14(–15)-locular; style ca. 20

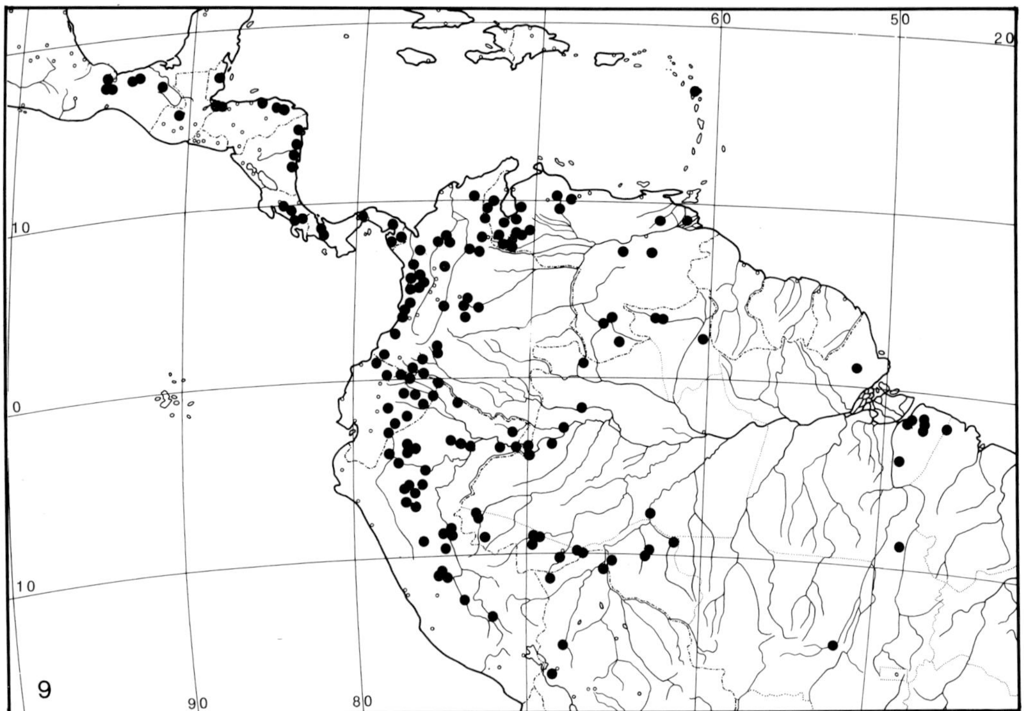

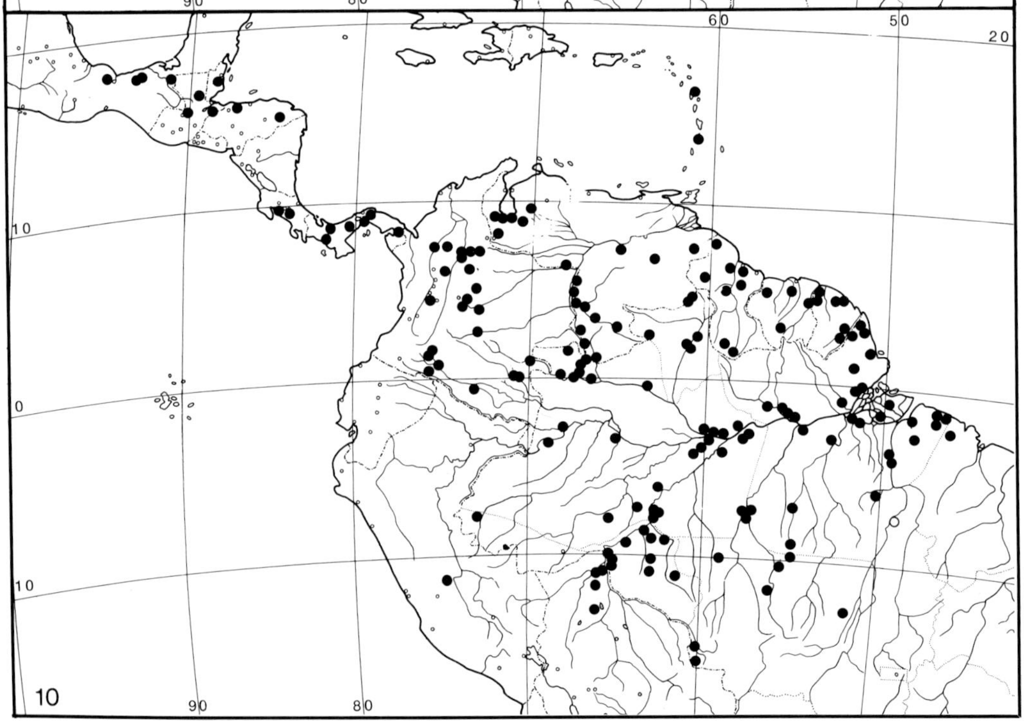

FIGS. 9, 10. Distribution of *Bellucia*. **9.** *B. pentamera*. **10.** *B. grossularioides*.

mm long. *Fruit* a greenish-yellow, subglabrous berry. *Seeds* ovoid, ca. 0.9 mm long, testa irregularly tuberculate and with shallow grooves.

Distribution and ecology (Fig. 10): *Bellucia grossularioides* is widespread in primary and secondary forest in Central America and northern South America. It grows on non-inundated soil, but also along river banks and in open, marshy places; from sea level to 850 m.

Representative specimens examined. MEXICO. CHIAPAS: Ruina, Palenque, *Matuda 3712* (NY). TABASCO: A la orilla del camino, km 50.3 de la desviación de Huimanguillo, hacia Fco. Rueda, *C. Cowan 2591* (CAS, NY). VERACRUZ: Ceiba Blanca, Mun. Las Choapas, *Dorantes et al. 2260* (CAS, US).

GUATEMALA. IZABAL: Izabal, *Bernoulli & Cario s.n.* (GOET). PETEN: Cadenas, E of km 169 of Poctun rd., *Contreras 5999* (DS, F, US). ALTA VERAPAZ: Cubilguitz, *Türckheim no. 4177, ser. I* (BR, G, NY, US).

BELIZE. CAYO: Hummingbird Hwy., ca. mi 40, S of Belmopar, *Dwyer 11354* (MO). STANN CREEK: Tohune ridge, Stann Creek–Capon rd., *Gentle 1940* (F, MO, NY).

HONDURAS. GRACIAS A DIOS: La Mosquita, nr. Río Patuca, nr. town of Wampusiripi, *Clewell 4510* (MO). YORO: Above Las Delicias, 20 km S of Tela, *Hazlett 3313* (F, US).

PANAMA. DARIEN: Periaque Hills, nr. Río Sabana and Lara, *Tyson et al. 4735* (MO). PANAMA: Gigante Bay, Barro Colorado Island, *Shattuck 1119* (F, MO, US).

GUADELOUPE. "Ad basis montium," *Duchassaing s.n.* (GOET, P); Sofaia, Ravine-Chaude, *Duss 3205* (NY, US).

MARTINIQUE. S. loc., "introduced from Guadeloupe," *Duss 1167* (NY, US).

COLOMBIA. AMAZONAS: Río Igara-Paraná, affl. Río Putumayo, corr. La Chorrera, terr. des indiens Witoto Jitomagaro, famille Kuiru, 17 km en avant de La Chorrera, *Gasche & Desplats 1187* (G, US). ANTIOQUIA: Vic. Planta Providencia, 26 km S and 23 km W (air) of Zaragoza, 7°13′N, 75°03′W, *Shephard 503* (COL, US). BOYACA: Orilla del cañon La Palmerena, 3 km al NE de Tauramena, *Uribe 4395* (COL, F, NY). CAQUETA: Florencia, *Cuatrecasas 8838* (COL, F, NY, US). CAUCA: Peña Blanca, *Woronow & Jazepczuk 4530* (NY). META: Puerto López, *Little & Little 8254* (COL, NY, US). PUTUMAYO: Río Putumayo en San Pedro, entre Umbria y Puerto Asís, *Cuatrecasas 10553* (COL, F, US). SANTANDER: La Gómez, 5 km W of Caserío, *Romero-Castañeda 6495* (COL, MO, NY). VAUPES: Nr. Mitú, *Cuatrecasas 6848* (COL, F, NY, US).

VENEZUELA. APURE: Distr. Camejo, bank of Río Meta, ca. 19 km (airline) WSW of Paraquito, 6°14′N, 68°05′W, *Davidse & González 13974* (MO, VEN). AMAZONAS: Pista de Sta. Bárbara del Orinoco, *Berry 715* (MO); southern part of Isla Raton, Orinoco, ca. 5°02′N, 67°46′W, *Breteler 4773* (F, G, IAN, MO, NY, U, US, VEN); Maroa, Río Guainía, *L. Williams 14342* (F, G, US, VEN). BARINAS: Socopo, Barinas–San Cristobal rd., *Marcano Berti & Peña 118-980* (BR, G, US). BOLIVAR: Entre Río Ikabaru y Uaiparu, *Bernardi 6581* (G, MO, NY); Bajo Caura, *Williams 11886* (F, NY, US, VEN). TACHIRA: Along Quebrada Sta. Teresa, just E of La Fria, 8°13′30″N, 72°14′30′W, *Steyermark et al. 120549* (MO, US, VEN). TRUJILLO: Cerro Gordo, cerca del pueblo de Cerro Gordo, 9°45′N, 70°15′W, *Steyermark & Carreño Espinoza 111657* (F, US). ZULIA: Nr. Campo Río de Oro, along Río de Oro, ca. 115 km WNW of Sta. Bárbara–San Carlos del Zulia, *Bruijn 1453* (MO, NY, US, VEN).

GUYANA. Northwest Distr., Koriabo river, Wauna Sub. Expt. Station, *Archer 2539* (NA, NY, US); Tumatumari, *Gleason 323* (BM, NY, US); Marudi Mts., along trail from Norman Mines camp to Aishalton, on abandoned airstrip, 02°15′N, 59°10′W, *Stoffers et al. 321* (B, U, US).

SURINAM. Zanderij, *Archer 2799* (NY, US); Saramacca river, *Maguire 23785* (BR, MO, NY, U, US, VEN); ca. 2 km below affl. of Oost river, Wilhelmina Mts., *Maguire et al. 54129* (F, IAN, MO, NY, U, US, W, VEN).

FRENCH GUIANA. Belizon, *Aubréville 298* (P, US); piste de St. Elie, *Granville B5122* (P); ca. 1 km S of mouth of Rio Yapoupi, 2°48′N, 52°28′W, *Irwin et al. 48449* (IAN, MG, MO, NY, RB, UB, US, VEN).

PERU. JUNIN: Km 41 on rd. to San Francisco de Satipo, 74°41′W, 11°10′S, *D. Smith et al. 1446* (MO, US).

BRAZIL. ACRE: Prox. do aeroporto de Cruzeiro do Sul, *Marinho 63* (IAN). AMAPÁ: Serra do Navío, 1°11′N, 52°08′W, *Pires et al. 51425* (F, IAN, MG, NY, SP, UB, U, US). AMAZONAS: Mun. São Paulo de Olivença, nr. Palmares, *Krukoff 8567* (B, BM, BR, F, G, MO, NY, U, US); Tefé, *Poeppig s.n. = 2528* (NY, W); Manaus–Porto Velho Hwy., km 235, *Prance et al. 20753* (HBG, INPA, MG, NY, S, U, US); São Gabriel, upper Rio Negro, along rd. to Camanaus, *Renner 365* (HBG); Manaus, Tarumã, *Renner 786* (HBG, INPA, US); 64 km N of Manaus on Manaus–Caracaraí rd., ZF-3 rd., km 33, *Renner 929* (INPA); vic. of Missão Salesiana, Rio Maturacá, *N. T. Silva & Brazão 60801* (F, MG, MO, NY, US); Rio Mapari, end of rd., ca. 30 km E of Borba, *Todzia et al. 2239* (NY, US). BAHÍA: In sylvis prope Iguarapi da Caxoeira, *Martius 2770* (M). MARANHÃO: Maracaçumé river region, *Fróes 1722* (BM, F, G, MO, NY, P, U, US); Anajatúba, *Ribeiro & Pinheiro 1211* (IAN, US). MATO GROSSO: 3 km W of Rio Suia–Missú Ferry, approx. 290 km N of Xavantina, *Ratter et al. 1421* (MO, NY, P, RB, UB, US); Boca do Rio Juruéna, *Pires 1480* (BR, P); inter Casalvasco et S. Luiz, *Riedel 1480* (BR, P); Aripuanã, BR 174, km 238, nucleo Juína, *M. G. Silva & Pinheiro 4299* (IAN, MG, NY). PARÁ: Altamira, Transamazônica, trecho Altamira-Itaitúba, km 23, Estação Exp. EMBRAPA, *Bahia 5* (MG, NY, US); Serra do Cachimbo, Cuiabá–Santarém rd., km 886, *Kirkbride & Lleras 2774* (BR, F, INPA, MG, MO, NY, US); Breves, Ilha de Marajó, *Pires et al. 5626* (IAN); Tucuruí, 12 km N of town along rd. to Cametá, ca. 3°36′S, 49°44′W, *Plowman et al. 9855* (MG, MO, NY, US); Santarém, Palhão rd., km 35, Igarapé Curupira, *M. G. Silva & Souza 2220* (G, MG, NY, US). RONDÔNIA: Mun. Sta. Barbara, BR

364, km 120, 12 km E da sede da mineração, 9°10′S, 63°07′W, *Teixeira et al. 811* (INPA); Rio Pacáas Novos, nr. first waterfall, *Ubiratan et al. 260* (IAN, INPA, MG, NY, US); nr. Tabajará, upper Machado river region, *Krukoff 1519* (BM, F, G, MO, NY, P, U, US). RORAIMA: Jarú, Rio Branco, *Kuhlmann 281* (RB, SP); Serra dos Surucucús, 2°42–47′N, 63°33–36′W, *Prance et al. 10086* (F, INPA, MG, NY, R).

BOLIVIA. BENI: 7–10 km SW of Guajará-Mirím on rd. to Riberalta, *Anderson 11919* (MO, NY, US); Prov. Vaca Diez, vic. Chácobo village Alto Ivon, 11°45′S, 66°02′W, *Boom 4285* (NY, US). PANDO: West bank of Río Madeira, 3 km above Abunã, *Prance et al. 8381* (F, INPA, MG, NY, P, U, US). LA PAZ: Prov. Iturralde, Luisita, 13°05′S, 67°15′W, *Beck & Haase 9999* (LBP, US).

Local names: capirote manzano (Nicaragua), néflier du Mexique (Guadeloupe), pomarrosa de monte, guayaba de pava, coronillo, manzanilla, takari (Ucbco), tuno, we-keé-to (Barasana), tsédjoó-ka (Punare), wach-keé-to (Tukano, all Colombia), puma rosa, pomarrosa, guayabo de danta, guaba de danta, cará-ré-yek (Pemon, all Venezuela), mess apple, chigonit, chiganet, itara (Arawak, all Guyana), nispa, mespelboom (Surinam), bois messe (Creole), melier, mêle (French Guiana), araçá de anta, goiaba de anta, papa terra branca, papa terra vermelha, jambo do mato, jambo da selva (all Brazil), huicama (Chácobo, Bolivia).

The scraped bark is rubbed on paddles, wooden implements, ropes, or drinking vessels to give a waterproof surface. Wood so treated has the appearance of being varnished. It has been suggested that jam could be made from the fruits (label data).

Bellucia grossularioides is illustrated in Seemann (1854), Mee (1968, pl. 7), and Cogniaux (1888, pl. 108, as *B. circumscissa*).

The species has been introduced in Guadeloupe (Fournet, 1978) and is sometimes cultivated in Martinique (fide *Duss 1167*). According to Grisebach (1864), *Bellucia grossularioides* has become naturalized in various southern Caribbean islands; however, I have seen no specimens other than from Guadeloupe and Martinique.

Near Manaus, in central Amazonia, *Bellucia grossularioides* hybridizes with *B. dichotoma.* I have collected putative hybrids along roadsides (*Renner 949, 957, 965, 969*) where they grow as vigorous, fertile trees. Pollen from these trees stains heavily in cotton blue/lactophenol and thus seems viable. In the size of the flowers and the number of petals, stamens, and ovary locules the putative hybrids are intermediate between the parents. For example, on one tree (*Renner 957*), I counted three flowers with 13 stamens, 12 with 14, four with 15, three with 16, and one with 18. Typically, *B. dichotoma* has 12 stamens and *B. grossularioides* 16. In the position of the cymes, the hybrids tend to resemble either the one parent or the other; in *Renner 965* the cymes were predominantly axillary, in *Renner 949* and *969* they developed mostly on the old wood. Depending on the position of the inflorescences, they are branched (on the old wood) or unbranched (in leaf axils).

5. **Bellucia acutata** Pilger, Verh. Bot. Vereins Prov. Brandenburg **47:** 181. 1905. Type: Brazil. Amazonas: Lower Marmelos, Mar 1902, *Ule 6097* (lectotype here designated: HBG!; isolectotypes: G!, MG!, F photo neg. 5976!). Holotype at B destroyed, but represented by a photo, F neg. 17248!. Figs. 1B; 2A; 11A, B.

Shrub or *treelet,* 1–6 m high; bark light gray, smooth. *Leaves* 11–15 × 4–6 cm, oblong (lanceolate), shortly acute, base rotundate or acute, the inner pair of lateral primaries departing 1–3 cm above the base, the outer pair very close to the leaf margin, glabrous; petioles 2.5 cm long. *Cymes* congested or rarely 2–3-branched, in the axils of existing leaves, glabrous; pedicels ca. 2 cm long. *Flowers* (6–)7-merous; hypanthium 4–5 mm long, glabrous; calyx a calyptra, dehiscing in an irregular semicircle, drying as a hyaline membrane, often persistent after anthesis; petals 12–15 mm long, white, pink-flushed outside; filaments 3–5 mm long, thecae ca. 5 mm long; ovary (8–)10(–11)-locular; style 10–13 mm long. *Fruit* a greenish-yellow, subglabrous berry. *Seeds* ovoid, ca. 0.7 mm long, testa irregularly tuberculate.

Distribution and ecology (Fig. 8): *Bellucia acutata* is characteristic of central Amazonian savannas, which are often subjected to regular burning and flooding.

Representative specimens examined. BRAZIL. AMAZONAS: Mun. de Nova Olinda, Rio Madeira, Posto da FUNAI, vila dos Indios Mundurucús, Rio Marimarí, *Cid 3987* (INPA, US); Mun. de Borba, Rio Abacaxis, lugar chamado Terra Preta, 04°22′S, 58°40′W, *Cid 4022* (INPA, US); Barcellos, Rio Negro, *Ducke 7155* (BM, MG); Humaitá, on plateau between Rio Livramento and Rio Ipixuna, *Krukoff 7093* (B, BM, BR, F, IAN, MO, NY, U, US); Humaitá, *Prance et al.*

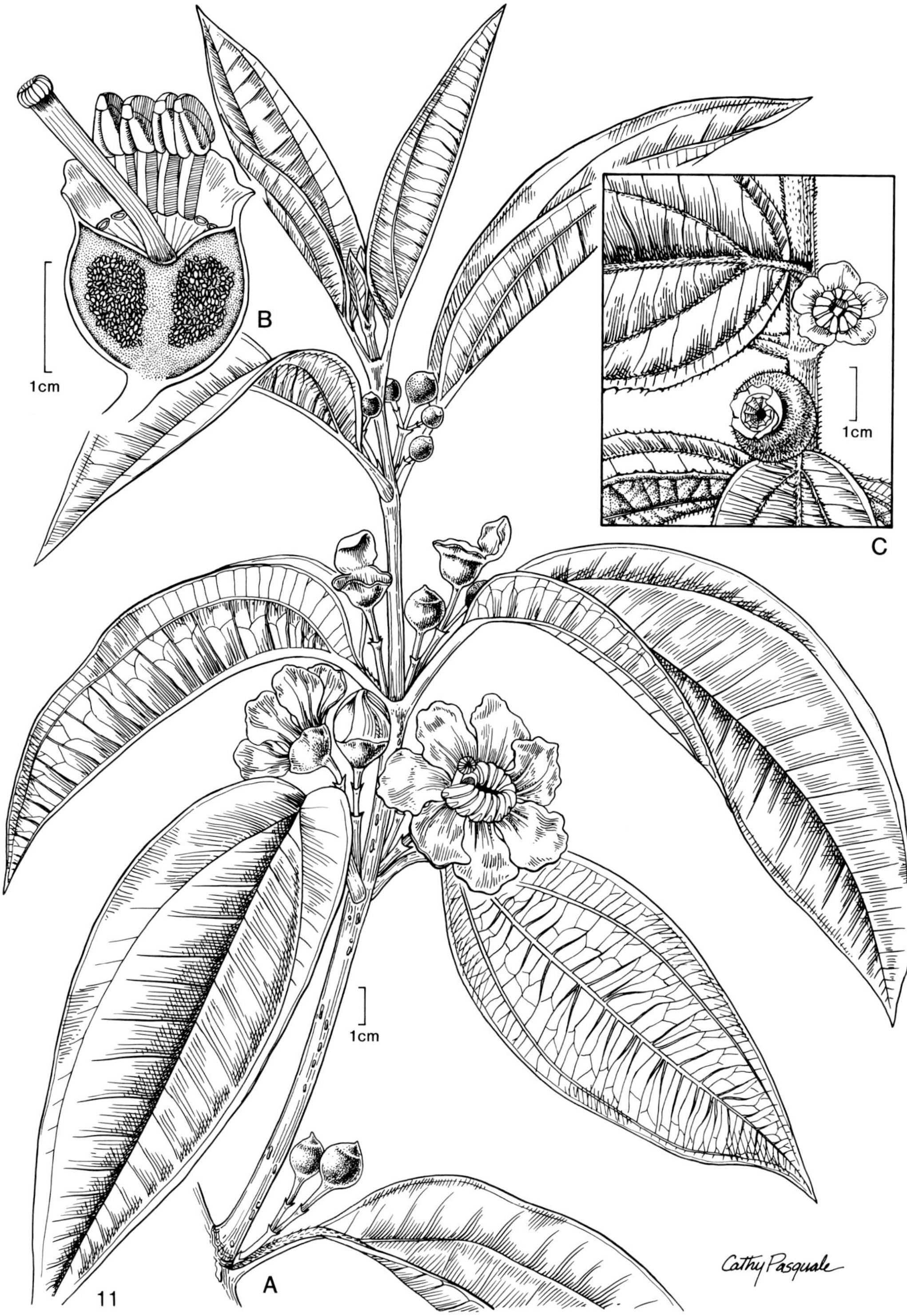

FIG. 11. **A, B.** *Bellucia acutata.* **A.** Habit. **B.** Longitudinal section through a flower, the petals and some of the anthers removed. **C.** Inflorescence of *B. beckii,* drawn from the holotype, *Beck & Haase 10000.*

3401 (COL, F, INPA, MG, NY, R, S, U, US), *Renner 533* (HBG, INPA); Fortaleza savanna, Rio Puciarí, trib. of Rio Ituxi, 20 km above mouth, *Prance et al. 13801* (COL, F, INPA, MG, NY, RB, SP, U, US); Cacau–Pirera, opposite Manaus, *Renner 337* (HBG, INPA, US); Manaus–Manacapurú rd., km 3, *Renner 959* (INPA, MG, UB). RONDÔNIA: Porto Velho, between km 8–23, *Maguire et al. 56703* (F, RB); Madeira–Marmoré railrd. nr. Abunã, km 220–223, *Prance et al. 5992* (COL, F, INPA, MG, NY, P, R, S, U, US); Porto Velho–Humaitá rd., km 75, *Prance et al. 19451* (INPA, MG, MO, NY, S, U, UFMT, US); Porto Velho, *Renner 987* (CAS, INPA, MG, UB); Rio Guaporé, margem do Rio das Garças, *N. T. Silva 378* (IAN, UB, US). RORAIMA: Rio Catrimani, Igarapé Camogi, *Pires 14055* (IAN, MG, US).

BOLIVIA. BENI: 7–12 km SE of Rio Yata on rd. from Guajará-Mirím to Cachuela Esperanza, *Anderson 11858* (MO, NY, US).

Bellucia acutata resembles *Bellucia grossularioides,* except for its lanceolate leaves, shrubby habitat, and small reproductive structures. It mostly has seven petals and 14 stamens, but only 10 ovary cells instead of 14 as would be the case in perfectly 7-merous flowers; occasionally, 11 or 12 cells develop in the ovary. This may be an indication that along with the evolution of the shrubby habit and axillary position of the flowers, an overall reduction in the size of the flowers and, concomittantly, the fruits occurred (compared to the ancestral condition which probably was *grossularioides*-like).

6. **Bellucia beckii** Renner, sp. nov. Type: Bolivia. La Paz: Prov. Iturralde, Luisita, 13°5′S, 67°15′W, pampa alta con islitas de pequeños árboles, 3 km W de Luisita, via Puesto Porvenir, 180 m, 25 Feb 1984 (fl), *Beck & Haase 10000* (holotype: US!; isotypes: LBP, US!). Fig. 11C.

A *B. acutata* ramulis villosis, foliis proportionaliter latioribus subtus sublanatis supra rugosis marginibus serrulatis, hypanthiis villosis vel sublanatis differt.

Shrub, 1–4 m high; twigs brown, sublanate, glabrescent. *Leaves* 16–18 × 8–11.5 cm, elliptic, shortly acute, rotundate at base, with inner pair of lateral primaries departing ca. 0.5 cm above the blade base, upper surface rugose, villose, glabrescent except on the central and lateral veins, lower surface villose to sublanate; petioles 1–1.5 cm long, villose. *Cymes* congested, in the axils of existing leaves, villose; pedicels ca. 0.5 cm long. *Flowers* 5-merous; hypanthium 5–6 mm long, villose to sublanate; calyx a calyptra, dehiscing irregularly, drying as a thin membrane, often persistent on the limb; petals 8–12 mm long, white; filaments 3–5 mm long, thecae ca. 5 mm long; ovary 9-locular; style 8–10 mm long. *Fruit* a greenish-yellow, loosely hirsute berry. *Seeds* ovoid, ca. 1 mm long, testa irregularly tuberculate.

Distribution and ecology (Fig. 8): *Bellucia beckii* occurs in savanna vegetation in northern Bolivia and in southern Mato Grosso and Pará (Brazil).

Additional specimens examined. BRAZIL. MATO GROSSO: Muns. Sinop & Colider, along BR 080, ca. 94 km E of jct. with BR 163, cerradão forest in the Serra Formosa range, 10°27′S, 54°08′W, 30 Oct 1985, *Thomas et al. 4165* (INPA, K, MG, NY, RB, SPF, US). PARÁ: Km 872 Cuiabá-Santarém; Serra do Cachimbo, 6 Nov 1977, *Prance et al. 24944* (MO, NY, RB, US).

BOLIVIA. Paratype: Topotypical, Sep 1985 (fl), *Haase s.n.* (US, GOET).

Bellucia beckii has characteristic leaves with a rugose upper surface and a softly pilose lower one and with serrulate margins; the twigs and hypanthia are covered with the same soft brown pubescence. The only other shrubby species of *Bellucia* with a calyptrate calyx, *B. acutata,* has smooth, glabrous leaves with entire margins. The berries of *B. beckii* are eaten by local people (R. Haase, pers. comm.).

7. **Bellucia huberi** (Wurdack) Renner, comb. nov.

Loreya huberi Wurdack, Phytologia **50:** 306. 1982. Type: Venezuela. Amazonas: At base of mountain north of Cerro Morrocoy ca. 10 km west of San Juan de Manapiare, basin of Río Manapiare, 5°19′N, 66°6′W, *Huber 1251* (holotype: US!; isotypes: US!, VEN!).

Shrub or *treelet,* 1–5 m high; bark light gray, smooth. *Leaves* (15–)19–25 × (6–)9–14.5 cm, broadly ovate, apically shortly acuminate, broadly acute or obtuse at base, with the inner pair of lateral primaries departing 1–1.5 cm above the base, glabrous; petioles (2–)3.5–4.5 cm long. *Cymes* congested, 1- to few-flowered, in the axils of existing leaves or cauline, glabrous; pedicels 5–10 mm long. *Flowers* 5–6-merous; hypanthium 3–4 mm long, apically glabrous, basally with few, inconspicuous hairs; calyx truncate, ca. 2 mm long; petals 12–12.5 mm long, white, pink-flushed outside; filaments 6–7 mm long, thecae ca. 5 mm long; ovary (7–)8(–9)-locular; style 8–10 mm long. *Fruit* a greenish-yellow, subglabrous berry. *Seeds* ovoid, ca. 0.8 mm long, testa irregularly tuberculate.

Distribution and ecology (Fig. 8): *Bellucia huberi* is known from savannas in the upper Río Orinoco area in Venezuela.

Additional specimens examined. VENEZUELA. AMAZONAS: Río Guainía, Maroa, *L. Williams 14387* (CAS, F, GH, MO, US, VEN), *L. Williams 14199* (F); Dept. Atures, valley of Río Coro-Coro, W of Serranía Yutajé, 5°41'N, 66°03'30"W, *Holst & Liesner 3230* (MO, US), *Liesner & Holst 21343* (MO, US).

The truncate calyx of *Bellucia huberi* is unusual within *Bellucia* and led to the original placement of this species in the genus *Loreya.* However, as noted in the protologue of *B. huberi,* the flowers of *Loreya* are usually considerably smaller. In the course of an investigation of the lower leaf surfaces of *Bellucia, Loreya,* and related genera at high magnifications, it became evident that *B. huberi* has the thick, papillose cuticle (Fig. 1G, H) characteristic of all species of *Bellucia* and absent in *Loreya.* Macroscopically, this causes the glaucescence of the lower leaf surface in *Bellucia huberi.* The upper surface is shiny and the texture is leathery due to the thick epidermal cuticle. *Loreya* leaves, on the other hand, have a completely smooth cuticle on both surfaces, which does not appear glaucous.

In number of floral parts, *Bellucia huberi* again does not fit in with *Loreya.* Of 14 flowers examined, seven had 12 filament scars, the others had 10, the number of ovary cells was seven (in two flowers), eight (in four), and nine (in one flower). On one flower, the stigma was preserved and had eight lobes. *Loreya,* on the other hand, is strictly pentamerous except for the rarely 6-merous species *L. subrotundifolia* and *L. riparia.* Also, *Bellucia huberi* is a shrub with sometimes axillary flowers, whereas all species of *Loreya* are trees with flowers borne on the old wood. Finally, *B. huberi* has seeds typical of *Bellucia* in shape and testa surface. In view of the extraordinary plasticity of the calyx in *Bellucia,* which has four species with calyptrate calyces and two with triangular sepal lobes, the truncate calyx in *B. huberi* cannot be given as much weight as hitherto. The papillose epidermis of the lower leaf surface, on the other hand, is likely to express a true phylogenetic relationship.

Excluded Taxa

Bellucia riparia Spruce ex Triana, Trans. Linn. Soc. London **28:** 142. 1871, nomen nudum.

Bellucia pluricaulis Spruce ex Triana, Trans. Linn. Soc. London **28:** 142. 1871, nomen nudum.

Ischyranthera laevigata Steudel, Pl. Surinam. Ed. R. F. Hohenacker, 1849, nomen nudum. Based on: *Kappler 2669,* distributed as *Hohenacker 1461.*

Bellucia arborescens (Aublet) Baillon, Hist. pl. **7:** 34. 1877.
[Basionym: *Melastoma arborescens* Aublet, Hist. pl. Guiane **1:** 420, tab. 163. 1775.]
=*Loreya arborescens* (Aublet) DC., Prodr. **3:** 179. 1828a.

Bellucia cacatin (Aublet) Sagot, Ann. Sci. Nat. Bot. VI, **15:** 332. 1883.
[Basionym: *Melastoma cacatin* Aublet, Hist. pl. Guiane **1:** 437, tab. 173. 1775.]
=*Miconia cacatin* (Aublet) Renner, comb. nov.

A recent collection from French Guiana (*Mori 18060*) has provided the first flowers of this troublesome taxon (compare Wurdack *in* Howard, 1983). I am indebted to John Wurdack who, after studying the flowers, suggested which of the ca. 1000 species of *Miconia* may be related to this species. Apparently, *Miconia cacatin* is in the lineage of sect. *Tamonea* somewhere between the species with 5-celled ovaries (*M. coronata* (Bonpl.) DC., *M. acuminifera* Triana) and the quite glabrous agglomeration with 3-celled ovaries starting in Cogniaux' (1891a) arrangement with *M. reducens* Triana.

Bellucia mespiloides (Miquel) Macbride, Field Mus. Nat. Hist., Bot. Ser. **13(4):** 498. 1941.
=*Loreya mespiloides* Miquel, Linnaea **18:** 619. 1844.

Bellucia spruceana (Triana) Macbride, Field Mus. Nat. Hist., Bot. Ser. **13(4):** 499. 1941.
=*Loreya spruceana* Triana, Trans. Linn. Soc. London **28:** 143. 1871.

Bellucia subrotundifolia Wurdack, Phytologia **45:** 331. 1980b.
=*Loreya subrotundifolia* (Wurdack) Renner, Mem. New York Bot. Gard. **50:** 46. 1988.

The reasoning behind the transfer of this species is discussed in my monograph of *Loreya*; briefly, *L. subrotundifolia* lacks the glaucous lower leaf surface of *Bellucia* and instead has the brown, smooth leaves of *Loreya*; it has (with occasional exceptions) 5-merous flowers with 5-lobed stigmas, a truncate calyx, pink flowers, purple fruits, and wedge-shaped seeds as well as other characters which make it a typical *Loreya.*

Bellucia umbellata Gleason, Bull. Torrey Bot. Club **58:** 257. 1931a.
=*Loreya umbellata* (Gleason) Wurdack, Phytologia **11:** 399. 1965.

Taxonomic Revision of *Loreya* (Miconieae)

INTRODUCTION

The neotropical genus *Loreya* De Candolle (Miconieae) comprises 13 species of trees and is centered in the southwestern part of the upper Amazon valley, though one widespread species reaches Nicaragua. The taxonomic circumscription of *Loreya* has long been problematic because the species vary with respect to the characters sometimes used to separate genera in Melastomataceae. Different species of *Loreya* have one or two anther pores, five or ten ovary cells, and branched or highly congested inflorescences. Moreover, leaf shape and venation, density and type of pubescence, and the size of the flowers vary strikingly within the genus. These difficulties notwithstanding, *Loreya* is geographically and ecologically coherent.

The need for an evaluation of *Loreya* arose during my monographic work on *Bellucia,* a genus that has sometimes been interpreted as synonymous with *Loreya* (Baillon, 1877; Macbride, 1941). In contrast to *Bellucia, Loreya* is poorly collected. The plants are medium-sized to fairly large trees occurring mostly in primary forest where they are infrequent. Only about 310 collections have been made since 1775, most of them after 1940. Naturally, the one treatment of the entire genus, by Cogniaux (1891a), is outdated. Moreover, among the more recently collected material from the upper Amazonian areas in Colombia, Ecuador, Peru, Bolivia, and Brazil are four new species. The present study attempts to clarify the generic limits of *Loreya* and place the recently collected novelties into a taxonomic framework.

TAXONOMIC HISTORY

The botanical history of *Loreya* began with a description and illustration of *Melastoma arborescens* by Aublet in 1775. The generic name *Loreya* was published by de Candolle (1828b) with *L. arborescens* the only cited species; the name honors the French scholar F. Lorey (?–1841), who wrote a flora of Burgundy and a checklist of the plants of the Ivory Coast. *Loreya,* as described by de Candolle, has a truncate calyx, strictly pentamerous flowers, and cymes on the old wood. Cogniaux (1891a) recognized seven species. Since then, an additional 19 species have been described. Macbride (1941), following Baillon (1877), synonymized *Loreya* under *Bellucia,* a related genus with similar floral morphology.

There have been a number of regional treatments of *Loreya* in the floras of Panama (Gleason, 1958), Ecuador (Wurdack, 1980a; see also Dodson & Gentry, 1978), Peru (Macbride, 1941), Venezuela (Wurdack, 1973), Guyana (Gleason, 1932), Surinam (Gleason, 1935), and French Guiana (Lemée, 1955).

VEGETATIVE ANATOMY

Here, I give a brief summary of Welle & Koek-Noorman's (1981) findings on the wood anatomy of *Loreya mespiloides, L. strigosa* (as *L. quadrifolia*), and *L. arborescens* (as *L. acutifolia* and *L. maguirei*).

Growth rings absent or present. Vessels diffuse, solitary and in radial multiples and irregular clusters of 2–7(8), 7–22(4–30) per mm^2 with alternate vestured pits. Vessel-ray and vessel-parenchyma pits variable, in part similar to the round and/or polygonal, sporadically oval intervascular pits, in part oblong and up to 30 × 6 μm, often coalescent. Fibers septate and sporadically non-septate as well, gelatinous in variable amounts. Fiber/vessel ratio 1.11–1.48. Parenchyma scanty paratracheal. Rays heterogeneous, uniseriate, occasionally with a biseriate part, 2–4 cells high, in the middle of the rays. The authors

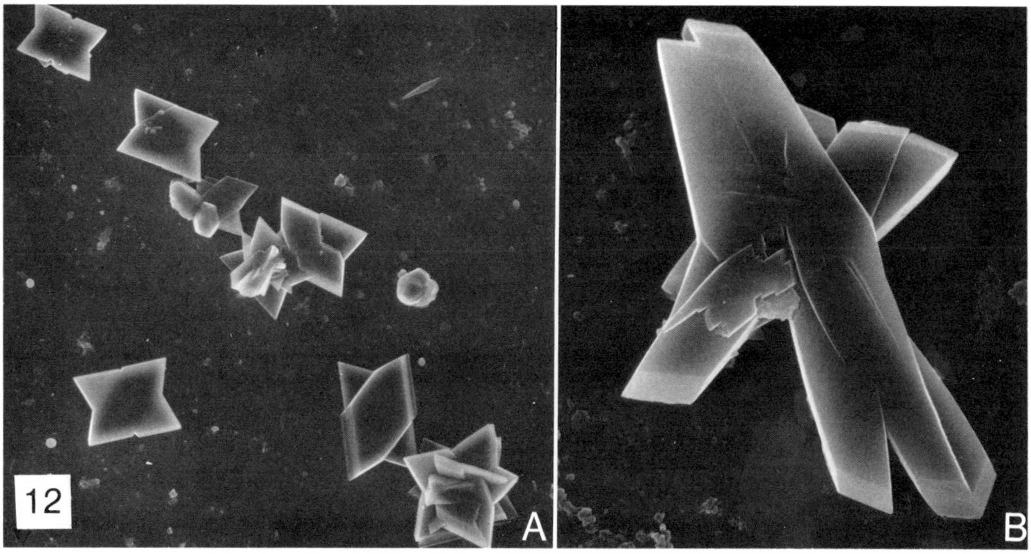

FIG. 12. Scanning electron micrographs of styloids of *Loreya riparia*. **A.** Crystal sand (×350). **B.** Crystal (×1500).

note that the fiber/vessel ratio in a sample of *L. arborescens* (as *L. acutifolia*) is very high (1.48) compared to the other species.

Loreya mucronata Gleason (=*Henriettea mucronata* (Gleason) Renner) had, "contrary to the general picture in *Loreya*, the bands of pseudoparenchyma consisting of fibers and parenchyma strands, the latter ones diffuse and in radial rows of up to 4 cells." I am excluding this species from *Loreya* on account of its morphological features (see EXCLUDED TAXA); the wood anatomical data support this decision.

The wood of *Loreya arborescens* and *L. mespiloides* is of intermediate weight (0.65–0.75 g/cm^3; Détienne & Jacquet, 1983), and light to dark brown. The bark on the branches and—as far as known—also on the trunk is generally smooth and thin (3–6 mm thick), the slash wound bright red or orange. An exception is *L. wurdackiana* in which the branchlets have longitudinally fissured bark.

The anatomy of the leaves of *Loreya spruceana* and *L. nigricans* was studied by Gottschall (1900; for summaries, see Solereder, 1908 and Metcalfe & Chalk, 1979) as part of his broad survey of the leaf anatomy of the Miconieae. He noted the homogeneous dark brown color of the dried leaves of both species, the smooth hairs of the first (which consist of a single row of cells), and the large styloids of the second. I have found styloids in *L. subrotundifolia*, *L. riparia* (Fig. 12), and *L. subandina* where they attain lengths of ca. 30 μm. These calcium oxalate crystals occur near the nerves and in at least some species just below the epidermis of the lower leaf surface.

Viewed by scanning electron microscopy (SEM) the epidermal cuticle of both leaf surfaces appears completely smooth in all species of *Loreya*.

MORPHOLOGY

Vegetative Structures

All *Loreya* species are trees. Most of them are medium-sized (5–10–15 m), some reach heights of 20 m or more (*L. arborescens*, *L. umbellata*, *L. gracilis*, *L. riparia*). *Loreya mespiloides* rarely reaches 18 m. The trunks are terete, generally smooth, and mostly of 10–12 cm DBH, although a diameter of 30 cm has been recorded for *L. gracilis*.

The tree architecture of *Loreya subrotundifolia* and *L. mespiloides* conforms to the Aubréville model in the classification of Hallé et al. (1978; Cremers, 1983). The form of growth is determined by a monopodial trunk with rhythmic growth and decussate phyllotaxis, bearing whorled tiers with similar phyllotaxis. Branches grow rhythmically, each branch plagiotropic by apposition. The inflorescences are lateral and the branches grow indefinitely.

The leaves present several taxonomically important characters since the length of the petiole, the shape of the lamina, and the pubescence vary considerably among the species. Generally, *Loreya* leaves are membranaceous or subcoriaceous and all have a completely smooth epidermal cuticle. They are elliptic, ovate to broadly ovate, or suborbiculate; *L. ovata* has a drip-tip. In some species the margin may be dentate, particularly in young leaves. Lower surfaces of leaves may be glabrous or moderately to densely strigose or sericeous. The hairs are simple, smooth, cylindrical, and white or reddish-brown. Short, yellowish, gland-tipped hairs are present on the lower leaf surface in *L. umbellata* and *L. gracilis.*

The venation is strongly 5(–7)-pli-nerved with the lateral primary veins diverging from the mid-vein 1–8.5 cm above the base. In *Loreya riparia* and *L. subrotundifolia,* the lateral primaries diverge almost at the blade base, which makes their leaves unusual. *Loreya nigricans* differs in having pinnately veined leaves, with the secondary veins leaving the mid-vein in a subopposite to irregularly alternate fashion.

Reproductive Structures

In *Loreya,* the *inflorescences* consist of a central axis bearing pairs of minute decussate bracts with cymose branches in the axils of the uppermost bracts. The inflorescence of *Loreya subrotundifolia* is analyzed and illustrated by Cremers in his study of the vegetative and reproductive architecture of Melastomataceae (1983, fig. 21,2, as *Bellucia subrotundifolia*). The same type of inflorescence is found in *Loreya riparia* and *L. arborescens.* In *L. gracilis* and *L. umbellata,* the branches are congested to the point of forming an umbel-like cluster. In both species, the inflorescences are supported by a long peduncle. The maximum number of flowers in such inflorescences is about 30.

In other species, the inflorescence consists of only a few cymose branches with 3–5 flowers supported by a very short peduncle; this is the case in *Loreya nigricans, L. ovata, L. wurdackiana,* and *L. subandina.* In *Loreya strigosa, L. mespiloides, L. spruceana,* and *L. klugii,* the inflorescences are sessile or subsessile clusters of flowers with several congested pairs of minute bracts at their base.

The inflorescences are always borne on older wood below the existing leaves, either on the branches or, more rarely, on the trunk.

Strict 5-mery is the usual condition in *Loreya flowers.* There are five petals, ten stamens, five ovary cells, and five lobes on the stigma. In two of the large-flowered species, *L. subrotundifolia* and *L. riparia,* 6-merous flowers may be present: of 30 flowers examined from one tree of *L. riparia,* 26 were pentamerous, three were 6-merous, and one was 7-merous. In *L. subrotundifolia,* Wurdack (1980b) found 26 5-merous and seven 6-merous flowers among 33 flowers from two specimens examined. In two other large-flowered species, *L. spruceana* and *L. subandina,* the perianth and androecium are 5-merous, but the number of ovary locules is ten.

The flowers of *Loreya* are about 2–3 cm in diam., with the exception of *L. gracilis*; it has much smaller flowers in which the petals are only 3 mm long. The hypanthium is hemispherical and puberulous, strigulose, or strigose. The calyx is unlobed or slightly lobed depending on the species. The pubescence on the inside of the calyx limb varies considerably within and between the species.

The petals, convolute in bud, are fleshy, and they often bear a transverse callus ridge (Figs. 19D, 21D) on the inside. There is little variability in petal color. In most species, the petals are pink in bud and pink on the outside during anthesis. On the inside, they are often paler pink, especially at the base, but there are no distinct color marks. Five species have purely white flowers.

The morphology of the stamens is homogeneous within the genus. A relatively thick filament supports a truncate anther of about the same length (Figs. 18E, 19H, 21E); both are laterally flattened and white, except for a dorsi-basal callus on the fleshy connective which is yellow. *Loreya gracilis* is exceptional: it has yellow stamens that turn red during anthesis. This condition is frequently encountered in the Miconieae, but as far as I am aware is absent elsewhere in the immediate *Loreya* alliance (*Bellucia, Henriettea, Henriettella, Myriaspora*). The apices of *Loreya* anthers are blunt, relatively fleshy, and almost as thick as the anthers. In most species, the anthers dehisce through two narrow, closely adjacent pores. In *L. klugii, L. mespiloides,* and *L. spruceana,* there is but a single pore (Fig. 21E). The rim of the pores is constricted.

The ovary apex and style are glabrous and the stigma is capitate and 5-lobed. Taxonomically, the most important feature of the gynoecium is the number of locules discussed above under the topic of flower mery. Sections through the ovary of *Loreya subandina* (*Wurdack 10178,* in the slide collection of Dr. R. Eyde at US) show that in this species the ten locules are of equal size, each with an axillary placenta.

Fruits in *Loreya* are berries containing thousands of small seeds (0.5–1.4 mm long). The berries are hemispherical, varying from 0.8 cm to 2 cm in diam., and at maturity are either pale yellow or else purple, bluish-white, blue, or black. The dark fruits are red when immature. There is a correlation between color, size, and pubescence. The yellow fruits, found in five species, are loosely strigose, strigulose, or puberulous and measure 1.2–2 cm in diameter. Fruits of the "red-turning-blue" type, found in eight species, are glabrous or glabrescent and attain diameters of (0.5–)0.8–1.2 cm. This correlation is broken in *L. strigosa,* which should have yellow fruits on the basis of its fruit size and pubescence, but is reported to have red fruits, turning dark blue when mature (*Prance et al. 13168*).

My survey, by light and scanning electron microscopy, of the *seeds* of *Loreya* (except *L. gracilis, L. nigricans,* and *L. strigosa*), *Bellucia, Henriettea,* and *Myriaspora* showed those of *Loreya* to be distinct from those of the other genera (except the monotypic *Myriaspora*; for differences see INTER- AND INTRAGENERIC RELATIONSHIPS) and of two types. Most *Loreya* species have wedge-shaped seeds, 0.9–1.4 mm long, with an irregular surface on which groups of adjacent testa cells form shallow grooves (Fig. 13D–F). The pronounced hilum is lateral and elongate. *Loreya klugii, L. mespiloides,* and *L. spruceana* have ovoid seeds, 0.4–0.5 mm long, with a regularly tuberculate surface formed by bulging periclinal cell walls (Fig. 13A–C). The hilum, always darker (brown or black) than the rest of the seed, is smaller and less pronounced in the ovoid seed type.

POLLEN MORPHOLOGY

I acetolyzed pollen of *Loreya riparia, L. spruceana,* and *L. subrotundifolia* according to standard procedures (Erdtman, 1960). Samples for SEM were first carbon-coated and then sputter-coated with gold-palladium and examined with a Hitachi S-570. The pollen grains (Fig. 14) of all three are radially symmetrical and isopolar; all have rugulate surfaces and all have 3-colporate apertures alternating with three subsidiary colpi.

POLLINATION AND DISPERSAL

Flowers of *Loreya* offer pollen as a reward for their bee pollinators. Odor has rarely been recorded in *Loreya,* and I have noticed no fragrance in *L. spruceana* or *L. riparia,* the floral biology of which I observed in the field (Renner, 1984a, 1986/1987). Vogel (1966) observed *Eulaema* bees visiting *L. spruceana.* The pollen of *Loreya* is also collected by species of *Xylocopa* and *Bombus* that vibrate the anthers and pollinate the flowers in the process. The bees land on the ring formed by the stamens (Figs. 18D, 19D, 21D) and vigorously shake them, causing pollen to shoot out of the anther pores. The pollen grains settle mostly on the lower side of the bees' bodies. Bees large enough to touch the stigma while collecting pollen potentially transfer pollen on contacting the sticky stigma of the next flower with their abdomen.

From label data and my own observations on *Loreya riparia* and *L. spruceana* near Manaus, Brazil, it appears that the species of *Loreya* have discrete, relatively brief (4–8 weeks) flowering periods. Fruit development begins immediately after anthesis and lasts one to two (to three?) months.

The diversity in fruit color and size suggests that an adaptive radiation in dispersal relationships has occurred within *Loreya.* The small, glabrous fruits, which are purple and turn blue or black when mature, are bird-dispersed. The larger, yellow, and more or less hairy fruits are eaten by such mammals as marsupials (Atramentowicz, 1982; Charles-Dominique, 1986), tapirs (label data: *Davis & Yost 937, Irvine 128*), and wild pigs (label data: *Davis & Yost 937*).

Even though a divergence within the genus into a mammal- and a bird-dispersed group has occurred, some species are associated with both dispersal agents. For example, the purplish-red fruits of *Loreya mespiloides* are removed diurnally and nocturnally at almost the same rate, being attractive to both marsupials and birds (Charles-Dominique, 1986). A further step might

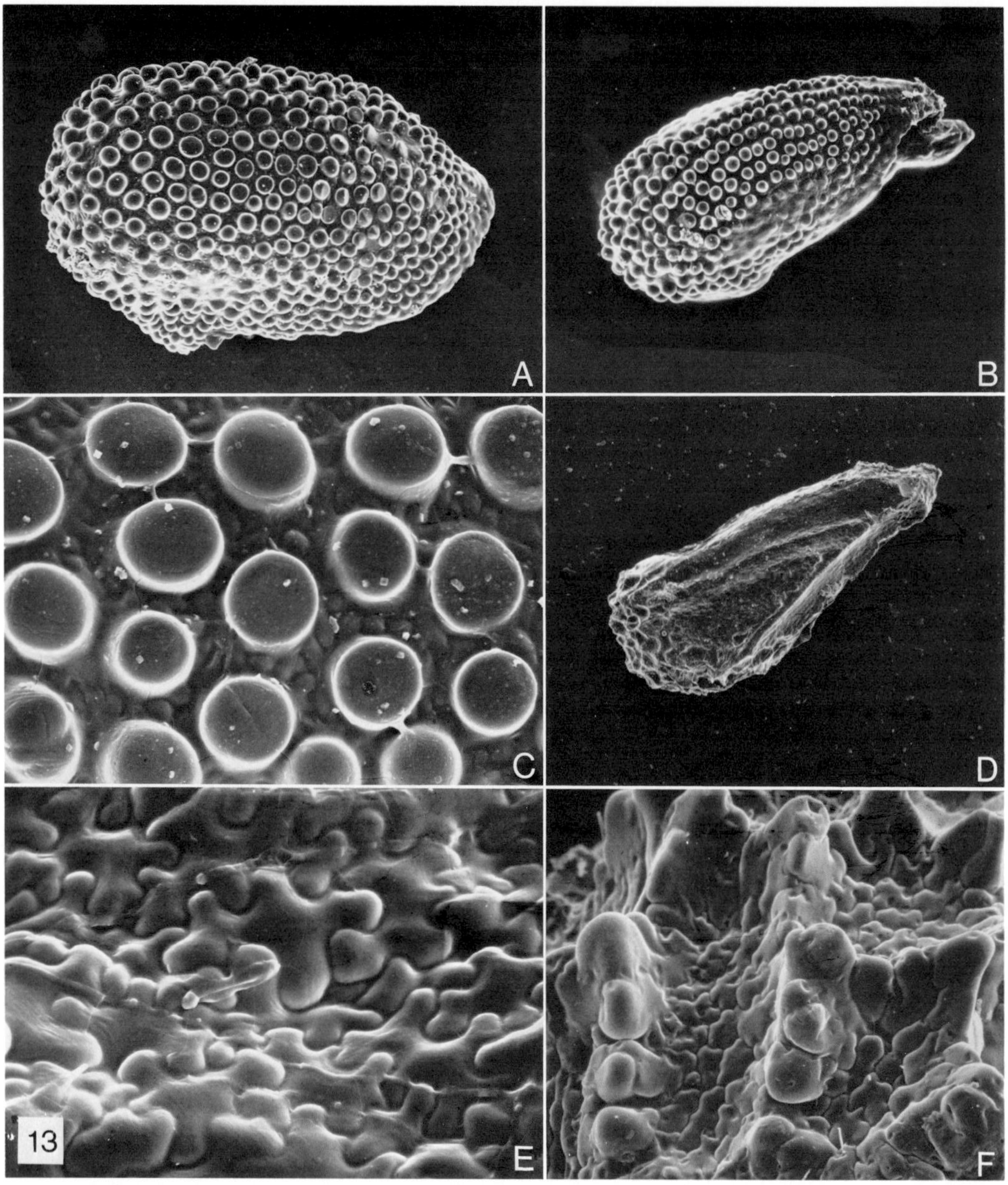

FIG. 13. Scanning electron micrographs of *Loreya* seeds and testa surfaces. **A.** *L. spruceana* (×100). **B.** *L. klugii* (×80). **C.** *L. spruceana* (×550). **D.** *L. umbellata* (×35). **E.** *L. subrotundifolia* (×550). **F.** *L. ovata* (×250).

be evidenced in the divergence seen between the species-pair *L. riparia* and *L. subrotundifolia.* These are morphologically similar; however, the first has larger, yellow and the second, smaller, purple fruits. (The two occur in different habitats and are geographically isolated; see DISTRIBUTION AND ECOLOGY.)

USES

Lemeé (1956) says that inhabitants of French Guiana eat the fruits of *Loreya arborescens* "avec plaisir," and label data show that fruits of *L. mespiloides* are also sometimes eaten. But the scarcity of common names (listed under the

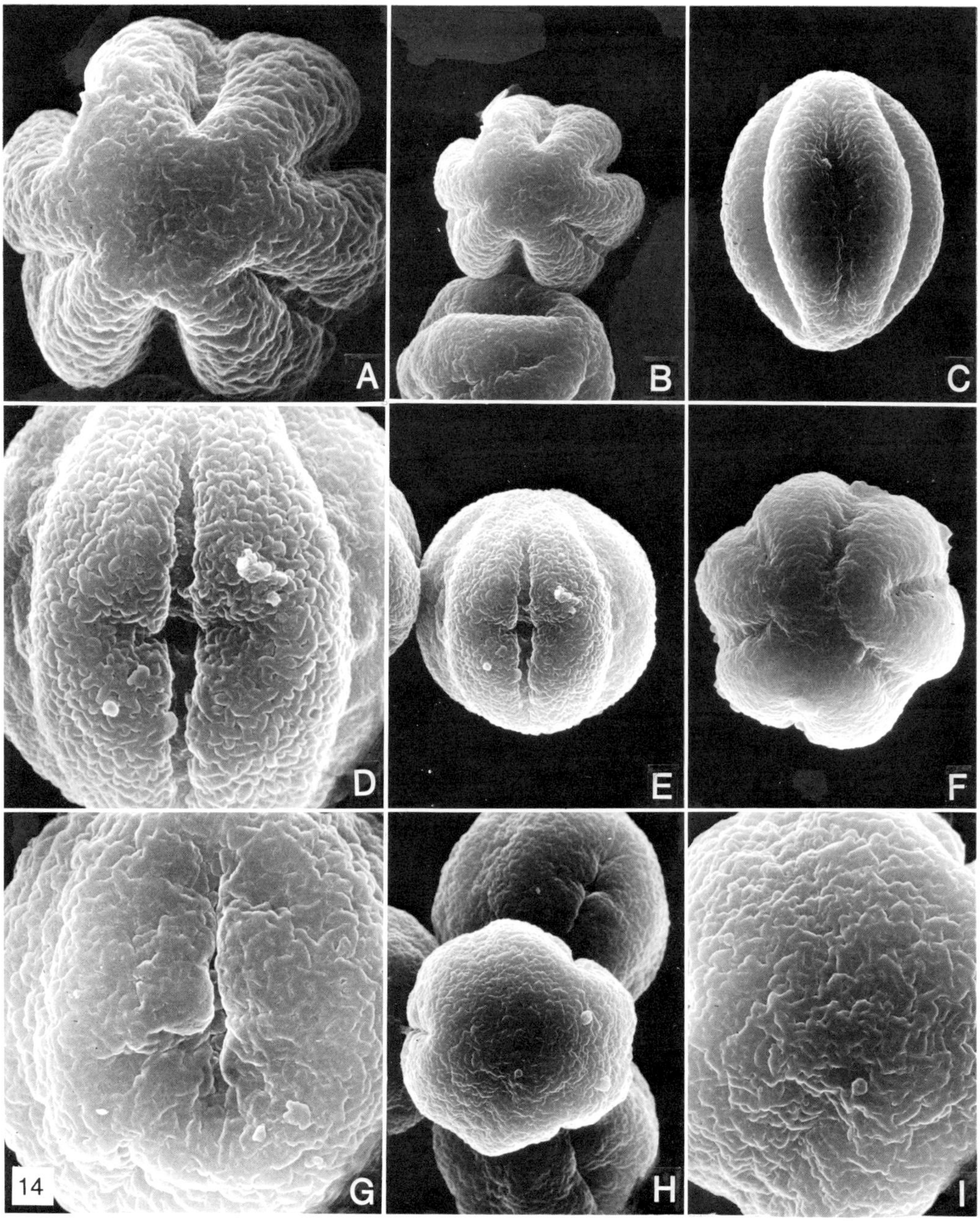

FIG. 14. Scanning electron micrographs of pollen grains in *Loreya*. **A–C.** *L. subrotundifolia*. **A, B.** Polar views (A ×5000, B ×2500). **C.** Lateral view (×2500). **D–F.** *L. riparia*. **D, E.** Lateral views (D ×500, E ×2500). **F.** Polar view (×2500). **G–I.** *L. spruceana*. **G.** Lateral view (×5000). **H, I.** Polar views (H ×2500, I ×5000).

species) indicates that the consumption of the fruits (and other uses) are of minor importance. The bark sap of *L. arborescens* is used as varnish by Arawak indians in Guyana.

Distribution and Ecology

Most *Loreya* species occur in the southwestern subandean Amazon basin (Figs. 15–17, 20). The most widespread species, *L. mespiloides*, occurs

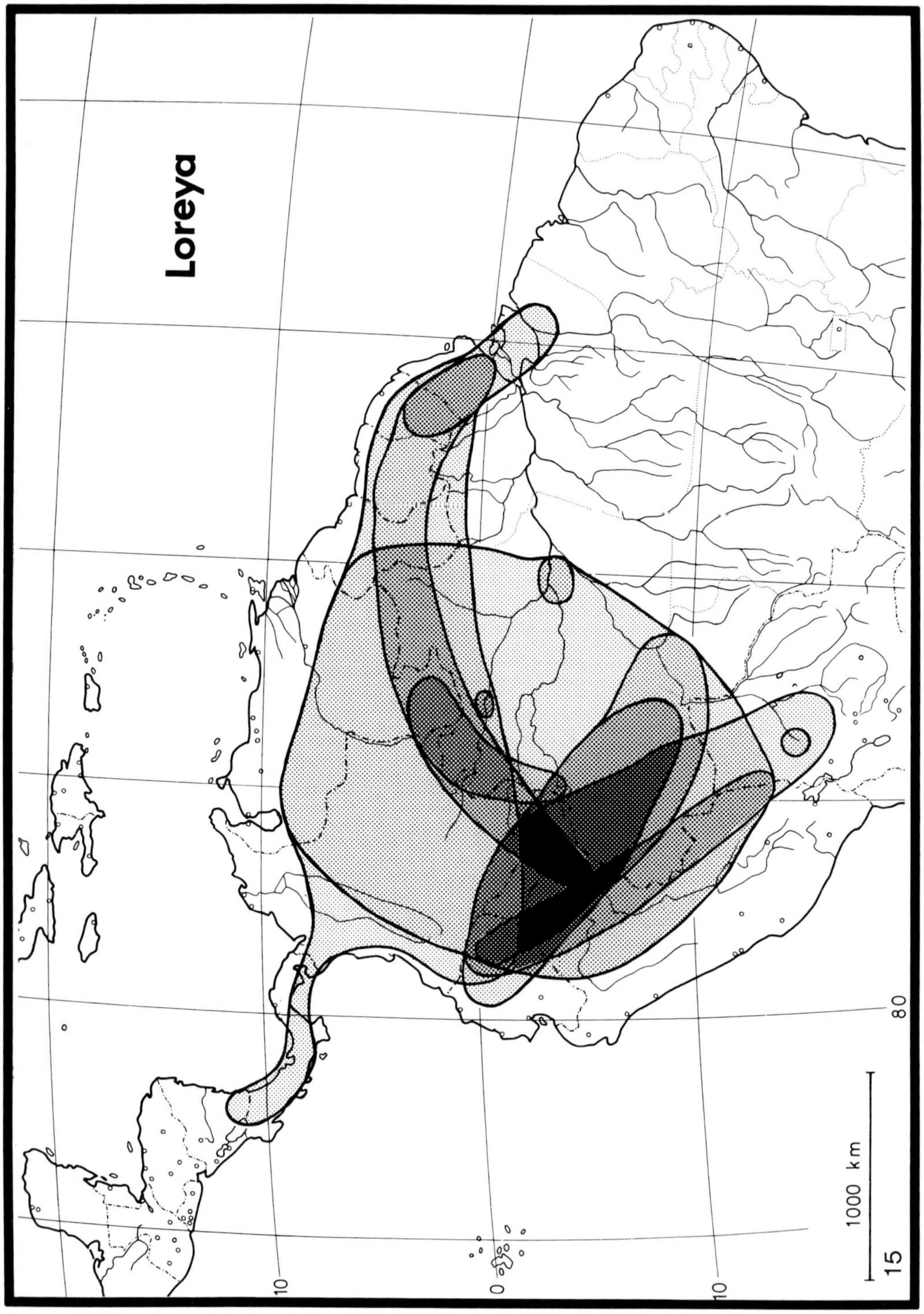

FIG. 15. Distribution of *Loreya*.

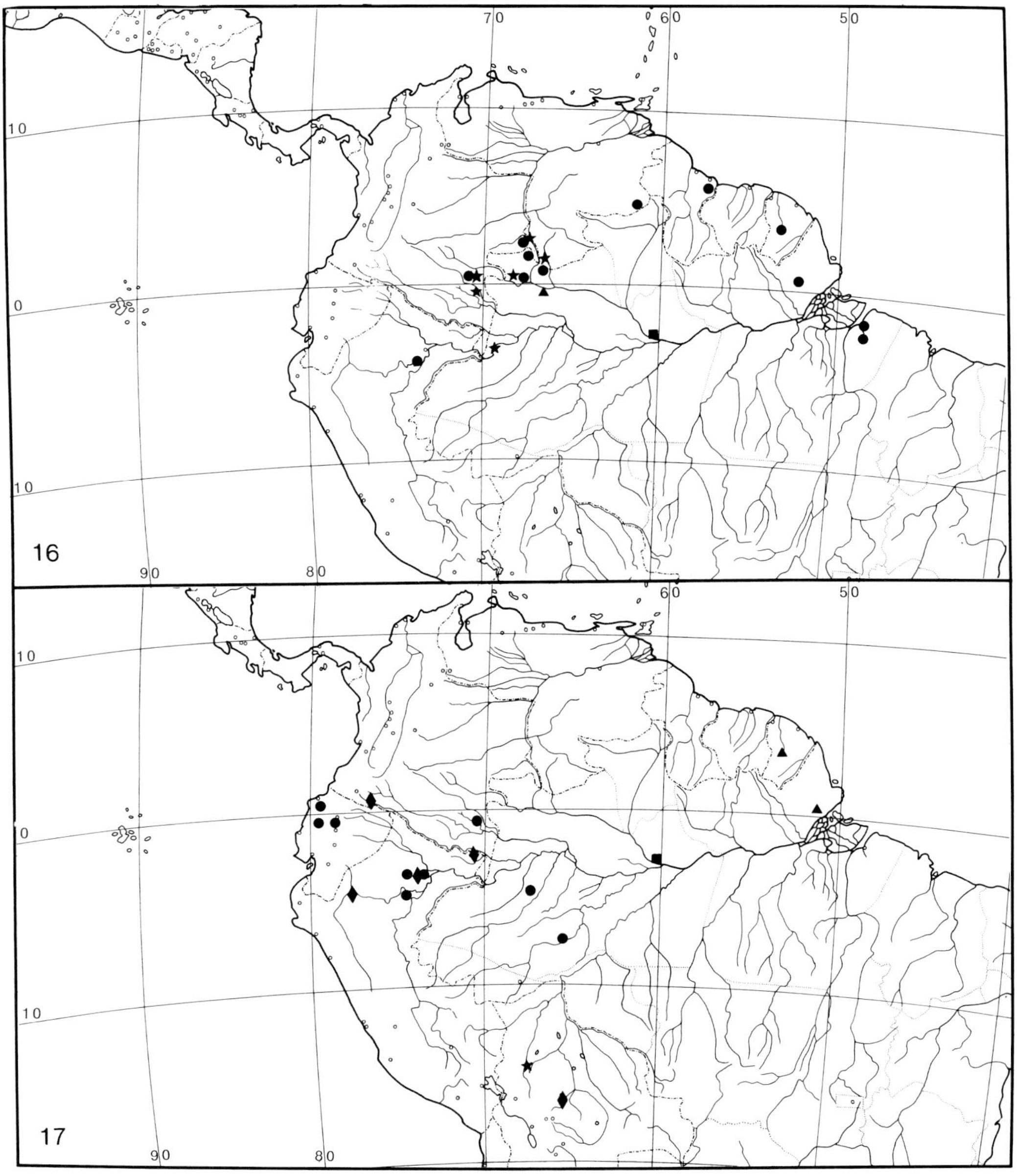

FIGS. 16, 17. Distribution of *Loreya*. **16.** *L. nigricans* (■), *L. wurdackiana* (▲), *L. ovata* (★), *L. arborescens* (●). **17.** *L. umbellata* (●), *L. gracilis* (★), *L. subrotundifolia* (▲), *L. riparia* (■), *L. klugii* (◆).

from Nicaragua (ca. 13°N) through Colombia, Ecuador, Peru, and Venezuela to the Guianas and Brazil. The southernmost species, *L. gracilis,* reaches ca. 15°S in the upper Río Beni region in Bolivia. Four of the 13 species appear to be local endemics, although future collections may extend their ranges; they are *Loreya wurdackiana* from a southern tributary of the upper Rio Negro, *L. nigricans* and *L. riparia* from the vicinity of Manaus, and *L. gracilis* from the Mapiri region in Bolivia.

Though *Loreya* species are mostly shade-loving trees of the rainforest, *L. mespiloides* achieves its greatest success in disturbed vegetation and

L. subrotundifolia is characteristically found in savanna habitats. *Loreya* is most diverse in the lowlands, but *L. arborescens, L. subandina, L. mespiloides,* and *L. spruceana* extend to 700–1200 m. *Loreya gracilis,* already mentioned above as the southernmost species, is so far only known from 850 to 1400 m altitude; it is also the smallest-flowered species.

Most species occur on well-drained upland clay soils ("terra firme"), though two species (*Loreya riparia, L. wurdackiana*) grow in periodically inundated areas adjacent to black water rivers ("igapó") where the soil is often sandy. The other riparian species, *L. umbellata, L. subandina,* and *L. spruceana,* grow on non-inundated river benches.

Loreya riparia, endemic to Manaus, and *L. subrotundifolia,* known from French Guiana and northeastern Brazil (Amapá), differ from each other mainly in the kinds of habitat they occupy (inundated forest vs. savanna) and the color, size, and pubescence of their fruits. Since *Loreya* is most diverse in the upper Amazon valley, it is tempting to hypothesize that the ancestor of *L. riparia* and *L. subrotundifolia* was some upper Amazonian species that spread to central Amazonia, the mouth of the Amazon and into French Guiana. Its range then became fragmented and two remaining isolates, one in the vicinity of Manaus and one at the mouth of the Amazon, subsequently diverged, entering new habitats and relying on different dispersers (as indicated by the fruit color, size, and pubescence; cf. POLLINATION AND DISPERSAL).

INTER- AND INTRAGENERIC RELATIONSHIPS

Loreya belongs in the Miconieae, one of two neotropical baccate-fruited tribes of the Melastomoideae. The Miconieae are distinguished from the other berry-fruited tribe, the Blakeeae, by their lack of persistent floral bracts. The Blakeeae also usually have a characteristic leaf venation, not found in Miconieae, in which the numerous secondary veins are so thin and run so close to each other that they almost look like striations.

Within the Miconieae, I consider *Loreya* to be most closely related to *Henriettella, Bellucia,* and *Myriaspora* and less so to *Henriettea. Henriettella* comprises 35–40 species of shrubs or small trees ranging from Guatemala and the West Indies (where it is most diverse) to Bolivia and southwestern Brazil. It has fascicles of sessile or short-pedicelled flowers borne on the branchlets below the extant leaves. The flowers are 4–5-merous, the anthers open by one or two pores, and the stigma is truncate rather than capitate as in *Loreya.* The leaves in *Henriettella* are 3–5-pli-nerved or 3-nerved, and often bear substellate or bullate hairs on the lower surface, a trait absent in *Loreya. Henriettella* differs further from *Loreya* in the morphology of its stamens, which have slender filaments and lack fleshy connectives. The diameter(s) of the pore(s) is (are) almost the same as the diameter of the anther in its upper section. *Loreya* anther pores, on the other hand, are small relative to the apex of the anther and have a constricted rim.

The genus *Henriettea* (some 12–15 species of shrubs or small trees occurring from Central America to the Amazon basin) doubtless belongs in this general alliance, but it is clearly distinct from *Loreya* in having purple, rostrate anthers and leaves that have bullate hairs on the upper surface. The seeds of seven species of *Henriettea* that I have looked at under SEM have a large lateral hilum surrounded by a protruding rim and on the surface bear long, tubular projections formed by the testa walls. They are unlike those of *Loreya.*

More problematic is the systematic treatment of the monotypic genus *Myriaspora* DC. (consisting of *M. egensis* DC. and including *M. decipiens* Naudin). *Loreya* and *Myriaspora* share several plesiomorphic characters, e.g., 5-merous flowers with 5-lobed stigmas and a truncate calyx. *Myriaspora* also has, in common with certain species of *Loreya,* at least three character states that I interpret as derived (I explain why below). These are 1-pored anthers, a 10-celled ovary, and seeds with a highly regular tubercular surface (as in *L. spruceana,* Fig. 1A). *Myriaspora* differs from *Loreya* in its shrubby habit and the position of the flowers in the axils of extant leaves rather than on older wood as in *Loreya.* This character distribution suggests that *Myriaspora* was derived from loreoid stock. The alternative, that *Myriaspora* and *Loreya* were derived from a common ancestor, is not supported by any shared derived characters.

Thus, the recognition of *Myriaspora* leaves *Loreya* paraphyletic, in a cladistic sense, since it does not include all its descendants. The recognition of *Bellucia* probably represents the same situation. The two genera have very similar flowers. However, this resemblance and the great intrageneric uniformity in floral morphology may be the result of a combination of shared ancestral traits and strong stabilizing selection by shared pollinators (Renner, 1986/1987). It cannot be the sole basis for judging their kinship. The synapomorphy that I have used to define *Bellucia* is the papillose cuticle on the lower surface. This papillose cuticle causes the glaucousness of the lower surfaces, whereas the leaves in *Loreya* are smooth and dark brown below.

None of the other characters are as persuasive. *Loreya* is strictly 5-merous with 5-lobed stigmas, except for occasional 6-merous flowers in two of its members (*L. subrotundifolia* and *L. riparia*). In contrast, *Bellucia* is 6–8-merous, except for *B. aequiloba,* which has five locules and 10 stigma lobes. The calyx of *Loreya* is a truncate limb with vestigial lobes; thus, the bud petals are exposed for many days before they expand. The calyx of *Bellucia* consists either of thick lobes or an irregularly rupturing calyptra, except in *B. huberi* where it is a truncate limb resembling that of *Loreya. Loreya* and *Bellucia* also differ with regard to seeds. *Bellucia* seeds are roughly ovoid with an almost terminal hilum. *Loreya* seeds are mostly wedge-shaped, and in three species where they are ovoid, the pattern formed by the bulging testa cells is much more regular than in *Bellucia.* Finally, in two species of *Loreya,* the form of growth follows the Aubréville model of Hallé et al. (1978; see VEGETATIVE STRUCTURES), whereas it approaches the Scarrone model in the four arborescent species of *Bellucia* (pers. obs.).

Given this distribution of characters, it seems most likely that *Bellucia* diverged from *Loreya,* possibly from the *subrotundifolia-riparia* lineage: this subgroup of *Loreya* is characterized by particularly large, occasionally 6-merous flowers, cymes of the same branching type as those of *B. aequiloba* and *B. dichotoma,* and—in one case—by yellow fruits like those of *Bellucia.* (However, yellow fruits arose several times within *Loreya*; see below.)

In my estimation, the maintenance of the single species of *Myriaspora* and of the seven species of *Bellucia* as separate genera from *Loreya* (which probably thereby becomes paraphyletic) is justifiable. Having names for these groups is useful because it recognizes that certain apomorphies such as those defining *Myriaspora* (shrubby habit, axillary flowers) and *Bellucia* (thicker epidermal cuticle, complete bud cover) have immense ecological consequences (reproduction at an earlier age, entering more open habitats where resistance to water stress is necessary, etc.). Both these genera are strikingly successful, weedy offshoots from *Loreya.* Also, there is the force of tradition and, in this case, tradition is reinforced by easy recognizability (*Loreya* leaves; *Bellucia* leaves) and phyletic unity of the groups.

Evolution within *Loreya* is more difficult to assess than the intergeneric relationships discussed above. Almost all the species are readily distinct on the basis of autapomorphies such as the pinnate venation in *L. nigricans,* the drip tip in *L. ovata,* the longitudinally fissured bark of the branches in *L. wurdackiana,* the small reproductive structures of *L. gracilis,* and the subsessile leaves in *L. strigosa.* However, the relations between the species, with few exceptions, are obscure because so few characters with clear polarity are useful in recognizing subgroups.

One of the few exceptions is the degree to which the cymes are condensed; developmental changes could have led to an increasing congestion of few-branched cymes and eventually to sessile clusters of flowers. Clustered flowers are found in *Loreya mespiloides, L. spruceana, L. klugii, L. subandina,* and *L. strigosa.* Four of these species share relatively large, hairy, yellow fruits. The fifth, *Loreya strigosa,* has red/blue fruits, which I consider the ancestral state because it is the rule in *Henriettella* and *Henriettea.* The four species sharing flower clusters and yellow fruits probably evolved the yellow color twice, independently, because three of them share 1-pored anthers and a regular tubercular testa sculpture (as in *L. spruceana,* Fig. 1A), whereas the fourth, *L. subandina,* has 2-pored anthers and wedge-shaped seeds. The one-pored state is derived from the two-pored one via merging of the pores. The regular testa surface pattern may be more specialized than an irregular one. The number of pores and the seed shape, in my opinion, are more likely to be true synapomorphies of *L. spruceana, L. mespiloides,* and *L. klugii* than the yellow fruit color and the reduction of the inflorescences which links them to *L. subandina.* On the other hand, *L. subandina* and *L. spruceana* are the only species of *Loreya* with 10-celled ovaries; the two also have very similar leaves. Thus, reliance on the number of anther pores and seed characters necessitates the postulation of an independent evolution of the 10-celled ovary in *L. subandina* and *L. spruceana*. However, such a multiplication in ovary cell number would seem to be a highly advantageous trait since it is correlated with an increase in flower and fruit size and thus in pollen and seed production. Therefore, given the capability in *Loreya* stock to increase the number of cells in the ovary, strong selection for the switch from five to ten locules was probably able to bring it about twice.

Yellow fruits evolved yet a third time in *Loreya riparia,* a species differing from other yellow-fruited species in that it has branched cymes and a distinctly different leaf shape and venation (also, two anther pores and five ovary locules). There is a tendency in *Loreya riparia* and *L. subrotundifolia* to produce 6-merous instead of 5-merous flowers; this capacity is relatively advanced since fixed flower mery is the ancestral condition in *Loreya.* The two species also share very broad, almost orbicular leaves in which the lateral primaries arise at the blade base. This seems derived because in the *Loreya* alliance pli-nerved leaves with the lateral primaries arising well above the base are the rule.

Further characters here considered derived, because they are unique features in *Loreya* and its relatives, are short glandular hairs on the lower leaf surface and umbel-like inflorescences borne on long peduncles. *Loreya gracilis* and *L. umbellata* share these characters and therefore appear to be each other's closest relatives. They differ in the size of the reproductive structures, which in *L. gracilis* are about half the size of those of *L. umbellata* (they are the smallest found in the genus). There is no obvious adaptive explanation for this reduction in size.

The remaining four species, *Loreya nigricans, L. arborescens, L. ovata,* and *L. wurdackiana,* are weakly united by the complete absence of any leaf pubescence. However, each of them is morphologically differentiated (cf. the above list of autapomorphies) to a degree that prevents postulations of their relationships. In *Loreya,* good characters, i.e., the number of anther pores and ovary locules, the type of inflorescence, the presence of glands on the lower leaf surface, and the seed morphology are few in number and beset with the possibility of parallelisms (see the example of the yellow fruits) and furthermore, in each case, unite only two or three species out of the 13. These circumstances prevent a meaningful cladistic analysis. Perhaps the abundance of autapomorphies and the scantiness of synapomorphies, the fact that the species seem isolated from each other, and the apparent absence of hybridization in *Loreya* are evidence of a long evolutionary history.

Taxonomic Treatment

Generic Description

Loreya De Candolle, Prodr. **3:** 178. 1828a. Type: *Loreya arborescens* (Aublet) De Candolle.

Heteroneuron Hooker f. *in* Bentham & Hooker, Gen. pl. **1(3):** 768. 1867.

Small to large *trees,* (3–)10–20(–38) m high; trunk cylindrical; bark smooth or rarely longitudinally fissured, light gray or brown, slash wound red. *Leaves* opposite, simple, petiolate, usually large and membranaceous, drying dark brown, loosely or densely strigulose, strigose, or sericeous with smooth eglandular hairs or rarely minute glandular hairs or glabrous, venation consisting of a mid-vein and 2–3 pairs of suprabasal primaries or venation pinnate (in *L. nigricans*), margin entire or serrulate (mostly in young leaves). *Inflorescences* borne on the old wood, few-branched cymes on a well developed peduncle, or contracted cymes with the peduncle short or lacking, or clusters of few flowers. *Flowers* hermaphroditic, 5-merous, except in two species where 6-merous flowers occasional; hypanthium hemispherical; calyx limb truncate, lobes vestigial; petals ovate, elliptic, or broadly triangular, apex obtuse, irregular, or subacute, fleshy, often with adaxial callus ridges and some development of lateral flaps, pink outside and pink and white inside, or completely pink, or white; stamens twice as many as the petals, isomorphic, glabrous, laterally compressed, the anthers and filaments of about equal length; anthers thick, 1- or 2-pored, the connective not prolonged but fleshy, with a more or less pronounced dorsi-basal callus; ovary completely inferior and fused with the hypanthium wall, 5(–6)- or 10-locular, apically with 10 radial ridges; stigma capitate and 5-lobed (occasionally 6-lobed in *L. riparia* and *L. subrotundifolia*); style glabrous, thick, terete, sigmoid, longer than the stamens. *Fruit* a many-seeded hemispherical, glabrous or hirsute, yellow or red/blue berry. *Seeds* minute, 0.5–1.4 mm long, ovoid or wedge-shaped with an elongate lateral hilum, testa regularly tuberculate or with an irregular surface of shallow grooves formed by groups of adjacent cells.

A neotropical genus of 13 species, ranging from Nicaragua to northern Bolivia, primarily in the upper Amazonian lowlands; lowlands to 1400 m.

Key to the Species of *Loreya*

1. Leaves pinnately veined. — 1. *L. nigricans.*
1. Leaves 5–7-pli-nerved.
  2. Anthers 1-pored.
    3. Leaf blades 6.5–9 cm wide, basally acute; hypanthium ca. 4 mm long; petals 6–7 mm long; stamens 7.5–10 mm long. — 13. *L. klugii.*
    3. Leaf blades 10–20 cm wide, basally obtuse or rounded to broadly acute; hypanthium (5–)8–10 mm long; petals 10–15 mm long; stamens 10–13 mm long.
      4. Leaves sericeous below; ovary 5-locular. — 11. *L. mespiloides.*
      4. Leaves puberulous to laxly strigulose below; ovary 10-locular. — 12. *L. spruceana.*
  2. Anthers 2-pored.
    5. Flowers in cymes, the peduncle 5–30 mm long.
      6. Cymes slender, 2–2.5 cm long; hypanthium ca. 2 mm long; petals ca. 3 mm long; stamens 4–5.2 mm long. — 6. *L. gracilis.*
      6. Cymes well developed, 3.5–5 cm long, hypanthium 3–10 mm long; petals 5–19 mm long; stamens 7–12 mm long.
        7. Leaf blades suborbicular or orbicular.
          8. Petioles (2–)3–4(–6) cm long; hypanthium 5–7 mm long; petals 13–15 mm long; ripe fruit purple. — 7. *L. subrotundifolia.*
          8. Petioles 8–10 cm long; hypanthium 9–10 mm long; petals 15–19 mm long; ripe fruit yellow. — 8. *L. riparia.*
        7. Leaf blades ovate, obovate, or elliptic.
          9. Lower leaf surface with short glandular hairs; cymes 3–4.5 cm long, umbellate, the peduncle ca. 3 cm long. — 5. *L. umbellata.*
          9. Lower leaf surface sparsely resinous-granulose; cymes 4.5–5 cm long, 2–3-branched (not umbellate), the peduncle 1.8–2 cm long. — 4. *L. arborescens.*
    5. Flowers in sessile or subsessile clusters, the peduncle short, to 5 mm long, or lacking.
      11. Leaf blades glabrous; the cymes once-branched, the peduncle short, 1–5 mm long.
        12. Leaf blades 17–23 × 11–15 cm, ovate to broadly elliptic, apex shortly acuminate to apiculate, without drip-tip. — 2. *L. wurdackiana.*
        12. Leaf blades 6.5–13.5 × 3–6 cm, elliptic, apex cuspidate, drip-tip 1–2 cm long. — 3. *L. ovata.*
      11. Leaf blades sparsely strigulose to densely sericeous; the flowers in sessile clusters.
        13. Leaves 7-pli-nerved, petioles 2–5 cm long; petals 15–18 mm long; stigma 2.5–3 mm in diam.; ovary 10-locular. — 10. *L. subandina.*
        13. Leaves 5-pli-nerved, petioles 0.5–0.8 cm long; petals 10–12 mm long; stigma 1.7–2 mm in diam.; ovary 5-locular. — 9. *L. strigosa.*

1. Loreya nigricans (Hooker f.) Triana, Trans. Linn. Soc. London **28:** 142. 1871.

Heteroneuron nigricans Hooker f. *in* Benth. & Hook., Gen. pl. **1(3):** 768. 1867. Type: Brazil. Amazonas: Manaus, Feb 1851, *Spruce 1331* (holotype: K!; isotypes: BR frag.!, G-BOISS!, M!, P!).

Tree, 5–10 m high. Branchlets and branches terete, with smooth, light gray bark. *Leaves* 9–15 × 4–7 cm, elliptic, shortly acuminate, obtuse at base, membranaceous, glabrous, the lower surface with needle-like styloids just below the cuticle parallel to the surface, venation pinnate; petioles 1–1.5 cm long. *Cymes* few-flowered, contracted, the peduncle to 5 mm long. *Flowers* with pedicels ca. 4 mm long; hypanthium 2–3 mm long, glabrous; petals white, 6–7 mm long, broadly ovate, apex irregular to emarginate; filaments 3–4 mm long; anthers 3–4 mm long, 2-pored; style 8–10 mm long; stigma ca. 1 mm diam.; ovary 5-locular. *Fruit* a glabrous, immature red, mature dark blue or black berry, ca. 0.8 cm diam. *Seeds* wedge-shaped with an elongate lateral hilum, bulging testa cells forming an irregular pattern.

Distribution and ecology (Fig. 16): *Loreya nigricans* is endemic to the Manaus area (Brazilian Amazonia), where it occurs on non-inundated soils in primary forest. It flowers in the latter part of the rainy season (Mar–Jul).

Representative specimens examined. BRAZIL. AMAZONAS: Manaus, *Ducke 171* (F, G, GH, IAN, MG, MO, NY, P, U, US), *Ducke 274* (IAN, MG, NY), *Fróes 20513* (IAN, MG, NY, US), *Pires 214* (IAN, MG, NY), *225* (IAN, MG, US), *Spruce s.n.* (NY), *Ule 5344* (GH, HBG, MG).

Loreya nigricans is distinct from all other species in having pinnately veined leaves. The species is illustrated in Hooker's *Icones plantarum* (1867).

Chenery (1948, and in a letter to J. Wurdack where the names of the species investigated are given) found that *Loreya nigricans* accumulates considerable quantities of aluminum in the leaves.

2. **Loreya wurdackiana** Renner, sp. nov. Type: Brazil. Amazonas: Rio Curicuriarí; mata baixa, solo arenoso, 26 Oct 1978, *O. C. Nascimento 761* (fl) (holotype: INPA!; isotypes: MG!, NY!, US!). Fig. 18.

Loreyae ovatae Berg ex Triana affinis, foliis majoribus proportionaliter latioribus pedicellisque longioribus differt.

Tree, 6–7 m high. Branchlets subangular, the bark brown and longitudinally fissured. Leaves 17–23 × 11–15 cm, ovate to broadly elliptic, shortly acuminate to apiculate, obtuse at base, membranaceous, glabrous and smooth, 5-plinerved, with the inner pair of lateral primaries diverging 2.5–3.5 cm above the blade base; petioles 3–5 cm long. *Cymes* few-flowered and contracted, the peduncle 1–2 mm long. *Flowers* with pedicels 10–15 mm long; hypanthium 2.5–3 × 4–5 mm, glabrous; petals white suffused with pink, 7–8 × 3–4 mm; ovate, apex slightly emarginate; filaments 4–5 mm long; anthers 3–3.5 mm long, 2-pored, connective dorsally thick and bulging; style 5–10 mm long; stigma ca. 1 mm diam.; ovary 5-locular. *Fruit* a glabrous, immature red, mature dark blue or black berry, ca. 0.8 cm diam. *Seeds* wedge-shaped with an elongate lateral hilum, testa cells forming an irregular pattern.

Distribution and ecology (Fig. 16): *Loreya wurdackiana* is only known from a southern tributary of the upper Rio Negro where it occurs in low forests on sandy soils which may be inundated for part of the year (igapó).

Additional specimens examined. BRAZIL. AMAZONAS: Along Rio Curicuriarí and Igarapé Cariúa to Cachoeira Piraiauará, 00°20′S, 66°55′W, *Alencar 599* (GH, INPA, MO, NY, US); along Rio Cariúa from Rio Curicuriarí to Cachoeira Boto, 00°20′S, 66°55′W, *Poole 2011* (INPA, MG, MO, NY, US); Rio Curicuriará, *Ducke s.n.* [=RB 25507] (RB).

This beautiful species is named as a tribute to Dr. John J. Wurdack who treated the Melastomataceae for the floras of Santa Catarina, Venezuela, Ecuador, and the Guianas.

3. **Loreya ovata** O. Berg ex Triana, Trans. Linn. Soc. London **28:** 142. 1871. Type: Brazil. Amazonas: Rio Uaupés, nr. Panuré, *Spruce 2899* (holotype: K!; isotypes: BR 2 sheets, one a frag.!, C!, F frag.!, G!, G-BOISS!, G-DC!, GH!, P!; isotype at B destroyed, but represented by a photo, F. neg. 17252!). Fig. 13F.

Loreya minor Cogniaux *in* Martius, Fl. bras. **14(4):** 522. 1888. Type: Brazil. Amazonas: Rio Sapo, Tocantins along Solimões, between Rio Jataí and São Paulo de Olivença, *Traill 311* (holotype: P!, F photo neg. 36363!; isotypes: BR frag.!, F frag.!, GH!, K!, P!).

Tree, 5–10 m high. Branches and branchlets terete, the bark smooth and light gray. *Leaves* 6.5–15 × 3–6 cm, elliptic, cuspidate, drip-tip 1–2 cm long, obtuse at base, subcoriaceous, glabrous above, sparsely puberulous and glabrescent below, 5-pli-nerved, the inner pair of lateral primaries diverging 0.4–0.7 cm above the blade base; petioles 1.5–2.5 cm long. *Cymes* few-flowered, contracted, the peduncle 2–5 mm long. *Flowers* with pedicels 4–5 mm long; hypanthium 3–4 × 5–6 mm, basally sparsely reddish-pulverulent, glabrescent; petals white suffused with pink, 8–10 × 3–4 mm, narrowly elliptic, apex irregular or emarginate; filaments 3–4 mm long; anthers ca. 4 mm long, 2-pored; style 8–10 mm long; stigma 1.5–1.7 mm diam.; ovary 5-locular. *Fruit* a glabrous, immature red, mature blue or black berry, ca. 0.8 cm diam. *Seeds* wedge-shaped with an elongate lateral hilum, bulging testa cells forming an irregular pattern.

Distribution and ecology (Fig. 16): *Loreya ovata* occurs in lowland rainforests on non-inundated ("terra firme") soils along the Rios Negro and Solimões.

Representative specimens examined. COLOMBIA. VAUPES: Río Inirida, 71°W, San Joaquin, *Fernández 1985* (COL, US); Río Apaporis, Jino-Goje (at mouth of Río Piraparana) and vic., general location 0°15′S, 70°30′W, *Schultes & Cabrera* 15659 (GH, US).

VENEZUELA. AMAZONAS: San Carlos de Río Negro, 1°56′N, 67°03′W, *Clark 7086* (NY, US); Yavitá, *L. Williams 13919* (F, G, NY, US, VEN).

FIG. 18. *Loreya wurdackiana.* **A.** Habit. **B.** Inflorescence, drawn from *O. C. Nascimento 761.* **C.** Fruits, drawn from *J. M. Poole 2011.* **D.** Flower. **E.** Stamens.

BRAZIL. AMAZONAS: Mun. São Paulo de Olivença, basin of creek Belém, *Krukoff 8708* (BR, F, G, GH, MO, NY, P, U, US); Equator, Rio Javarí, *Prance et al. 23941* (GH, INPA, M, MG, NY, U, US).

Local names and uses: palo de escoba (Venezuela), which would indicate that the wood is used for some household items such as brooms and sticks.

Loreya ovata is illustrated in Cogniaux (1888, pl. 110); its staminal vessel traces have been analyzed by Wilson (1950, sub *L. minor*).

Examination of the types of *L. minor* and *L. ovata* in Paris and of modern material showed that the leaf shape and stigma characters used by Cogniaux to distinguish two entities do not hold.

4. Loreya arborescens (Aublet) DC., Prodr. **3:** 179. 1828a. *Melastoma arborescens* Aublet, Hist. pl. Guiane **1:** 420, pl. 163. 1775; *Bellucia arborescens* (Aublet) Baillon, Hist. pl. **7:** 34. 1877. Type: French Guiana. Nr. Aroura, *Aublet s.n.* (holotype: BM!; isotypes: P frag.!, LINN-SM, fide Howard, 1983).

Loreya acutifolia O. Berg ex Triana, Trans. Linn. Soc. London **28:** 142. 1871. Type: Venezuela. Amazonas: San Carlos, *Spruce 3085* (lectotype, here designated: K!; isolectotypes: BR!, G!, G-BOISS!, G-DC!, GH!, NY!, P!).

Loreya maguirei Wurdack, Phytologia **18:** 160. 1969. Type: Venezuela. Amazonas: Río Cuao, *Maguire & Politi 28164* (holotype: NY!; isotype: US!).

Tree, 10–25(–38) m high. Branchlets slightly angular or sulcate, bark dark gray or brown, longitudinally fissured. *Leaves* 9–15 × 4–8(–12) cm, ovate, elliptic, or obovate, acuminate, bluntly acute or rounded, cuneate at base, membranaceous to subcoriaceous, glabrous on both surfaces, puberulous on the veins below, 5-plinerved, the inner pair of lateral primaries arising at the blade base or diverging 5 mm above the base; petioles 1–3 cm long. *Cymes* 2–3-branched, 4.5–5 cm long, the peduncle 18–20 mm long. *Flowers* with pedicels 5–10 mm long; hypanthium 3–4 mm long, basally reddish puberulous; petals pink outside and white within, 5–6(–7) × 3.7–4 mm, ovate, apex irregular or emarginate; filaments 3–4.5 mm long; anthers 3–4.5 mm long, 2-pored; style 9–15 mm long; stigma 1–2 mm diam.; ovary 5-locular. *Fruit* a puberulous, glabrescent, immature red, mature light bluish-white or blue berry, ca. 0.8 cm diam. *Seeds* wedge-shaped with an elongate lateral hilum, bulging testa cells forming an irregular pattern.

Distribution and ecology (Fig. 16): *Loreya arborescens* occurs in the northern region of the Amazon valley from Colombia to northeastern Brazil in primary and secondary vegetation at elevations from 140 to 1200 m. The species probably occurs in Surinam but has not yet been collected there.

Representative specimens examined. COLOMBIA. VAUPES: Río Apaporis, between Río Kananari and Río Pacoa, *García-Barriga 14066* (COL, NY, US); Río Negro, at confluence of Ríos Guainía and Casiquiare, Cañon Ducuruapo (Igarapé Rana), *Schultes & López 9360* (NY, US); Río Vaupes, Mitú and vic., *Schultes et al. 24417* (GH, US).

VENEZUELA. AMAZONAS: Maroa, 2°43′N, 67°38′W. *Clark 6866* (MO, NY, VEN); 1–5 km SE of Maroa, *Maguire et al. 41778* (B, COL, F, G, NY, P, U, US). BOLIVAR: El Dorado–Sta. Elena de Uairén rd., 14 km from Piedra de la Vírgen, *Marcano-Berti et al. 50981* (U, US).

GUYANA. MAZARUNI POTARO: Mazaruni river branch, Bartica–Kaburi rd., *Hohenkerk 874* (NY); Makauria Creek, *FD 3314/field no. 578* (NY).

FRENCH GUIANA. Saint-Elie, *Foresta 637* (U, US).

PERU. LORETO: Portero de Jenaro Herrera, right bank of Río Ucayali, *Encarnación 26250* (MO, NY, US).

BRAZIL. AMAZONAS: Cucuí, *Ducke s.n.* = *RB 41350* (NY); Rio Uaupés, Taraquá, *Pires 904* (IAN, NY); Panuré, *Spruce 2588* (BR, BR frag., C, G, G-BOISS, GH, K, MG, NY, P; B destroyed, but represented by a photo, F neg. 17251). PARÁ: Belém, *Ducke 1720* (IAN, MG, NY, US); Road BR 22, Capanema to Maranhão, km 96, vic. of Cachoeira, *Prance & Pennington 1839* (F, G, GH, IAN, MG, NY, U, US).

Local names and uses: trompo de lomorado (Simure tribe in Venezuela); itarra (Arawak, Guyana). The bark sap is used as varnish by the Arawak indians.

Loreya arborescens is illustrated in Aublet (1775, pl. 163) and in Cogniaux (1888, pl. 110, as *L. acutifolia*).

The species exhibits considerable variation in the shape of the leaves and the dorsal thickening of the connective. Two entities have been named, *L. acutifolia,* characterized by acute leaves and a pronounced dorsi-basal thickening of the connectives, and *L. arborescens,* the type of which has rounded leaf apices and lacks conspicuously thickened connectives. In addition, *L. acutifolia* was described as differing in the truncate or hardly enlarged stigma. This observation must have been made on a stigma that was either damaged or very young. Fully mature styles of specimens matching the syntypes of *L. acutifolia* have the typical capitate, 5-lobed *Loreya* stigma. As far as concerns the other alleged differences, specimens from French Guiana have obtuse leaf apices and hardly enlarged connectives, those from Guyana have similar connectives but acute leaf apices, and specimens from the more western part of the range of *L. arborescens* also have acute leaves, but have thickly developed connectives. Thus, the characters thought to distinguish two entities within *L. arborescens* are combined in plants from Guyana.

The type collection of *Loreya maguirei* differs in the inconspicuous outer pair of lateral primaries from the distinctly 5-nerved leaves of *L. arborescens* (incl. *L. acutifolia*). Similar venation is found in the collections *Ducke 1720* and *Clark 6866,* both typical of *L. arborescens* in all other respects.

Chenery (1948, and in a letter to J. Wurdack where he gives the names of the species he investigated) has found that *Loreya arborescens* (under *L. acutifolia*) accumulates considerable quantities of aluminum in the leaves.

5. Loreya umbellata (Gleason) Wurdack, Phytologia **11:** 399. 1965. Fig. 13D.

Bellucia umbellata Gleason, Bull. Torrey Bot. Club **58:** 257. 1931a. Type: Peru. Loreto: Mishuyacu, nr. Iquitos, *Klug 131* (holotype: NY!; isotypes: BM!, F!).

Tree (5–)8–13(–20) m high. Branchlets subangular, with a velvety brown hair-covering; bark dark brown, smooth. *Leaves* 25–30(–40) × 17–15(–25) cm, ovate, elliptic, or obovate, shortly acuminate, obtuse to rounded at base, membranaceous, adult leaves on the surface above glabrous, below with short glandular hairs, strigulose on the veins, 5-pli-nerved with the inner pair of lateral primaries diverging 5–6 cm above the blade base; petioles (2–)3(–5) cm long. *Cymes* many-flowered with 20–30 flowers, umbellate, 3.0–4.5 cm long, the peduncle ca. 3 cm long. *Flowers* with pedicels 8–10 mm long; hypanthium 4–5 mm long, deciduously puberulous; petals white to pink, 6–8(–9) mm long, 4 mm wide at the base, broadly triangular, apex subacute, filaments 5.0–5.5 mm long; anthers 5.0–5.5 mm long, 2-pored; style 14–16 mm long; stigma 1.8–2.0 mm diam.; ovary 5-locular. *Fruit* a puberulous, glabrescent, immature red, mature blue or black berry, ca. 0.8–1 cm diam. *Seeds* wedge-shaped with an elongate lateral hilum, bulging testa cells forming an irregular pattern.

Distribution and ecology (Fig. 17): *Loreya umbellata* is a riparian species, widespread in Ecuador and occurring in the Amazonian parts of Colombia, Peru, and western Brazil.

Representative specimens examined. COLOMBIA. VAUPES: Río Apaporis, Jino-Goje; entre los Ríos Piraparana y Popeyaka, *García-Barriga 14420* (COL, US).

ECUADOR. ESMERALDAS: Quinindé, about 3 km E, *Little 6223* (NY, US). LOS RIOS: Río Palenque Biological Station, km 56 Quevedo–St. Domingo, *Dodson 5826* (MO, QCA, US). PICHINCHA: 20 km W of St. Domingo de los Colorados, *Cazalet & Pennington 5049* (NY, US).

PERU. LORETO: Vic. of Lago Llanchama nr. Río Nanay, *Croat 18716* (AAU, MO, NY, U, US); cerca de Zungaro Cocha, 15 km S de Iquitos, *Dodson 2807* (MO, US); Jenaro Herrera, Río Ucayali, km 5 carretera Jenaro Herrera–Puerto Angamos, 73°40′36″W, 4°55′18″S, *Mejía 320* (F, US).

BRAZIL. AMAZONAS: Jamamadí Indian village, W of Rio Purús, 65°17′W, 7°28′S, *Prance et al. 23439* (INPA, MG, NY, U, US); Carauarí, NE of Rio Juruá, *A. S. L. da Silva et al. 404* (MG, US).

Local names: rifarillo (Peru); this name is also used for *Loreya klugii.*

Loreya umbellata and *Loreya gracilis,* the following species, are readily distinguished from all other species by the short, yellow, glandular hairs on the lower leaf surface and by the umbel-like inflorescences on long peduncles. Though not found elsewhere in *Loreya,* this type of short glandular hairs is widespread in the Miconieae (Wurdack, 1986). In the Flora of Peru (Macbride, 1941), the key gives *L. umbellata* as having anthers one-pored and *L. spruceana* as two-pored (the reverse of the actual case).

Loreya umbellata is illustrated in Dodson and Gentry (1978).

6. Loreya gracilis Renner, sp. nov. Type: Bolivia. La Paz: Prov. Nor. Yungas, 10 km by rd. (ca. 5 km by air) N and above Caranavi, ca. 15°47′S, 67°32′W, alt. 1400 m, 1 Nov 1984 (fl), *Nee & Solomon 30308* (holotype: US!; isotypes: F!, K!, LPB, MO!, NY!, P!).

Loreyae umbellatae (Gleason) Wurdack affinis, hypanthiis petalisque proportionaliter manifeste minoribus differt.

Tree, (8–)12(–20) m high, to 30 cm in diam.

Branchlets angular, with a brown, velvety hair-covering; bark brown, longitudinally fissured. *Leaves* 16–17 × 7–8 cm, ovate-elliptic, shortly acuminate, obtuse at base, membranaceous, glabrous above, lamina, mid-vein, and lateral primary and secondary veins with minute glandular hairs beneath, 5(–7)-pli-nerved, the inner pair of lateral primaries diverging 2–2.5 cm above the blade base; petioles 2–3 cm long. *Cymes* many-flowered, umbellate, delicate, 2–2.5 cm long, the peduncle 1–1.2 cm long. *Flowers* with pedicels 1.5–2 mm long; hypanthium ca. 2 × 1.2 mm, basally puberulous; petals white, ca. 3 mm long, basally ca. 3 mm wide, triangular, apex subacute; filaments 2–3 mm long; anthers 2–2.2 mm long, 2-pored; style 4.8–5.0 mm long; stigma 0.5–0.8 mm diam.; ovary 5-locular. Ripe *fruit* unknown, probably a red to black berry, ca. 0.5 cm diam. *Seeds* wedge-shaped with an elongate lateral hilum, bulging testa cells forming an irregular pattern.

Distribution and ecology (Fig. 17): *Loreya gracilis* is only known from the Mapiri region in Bolivia at 850–1400 m elevation.

Additional material examined. BOLIVIA. LA PAZ: Prov. Larecaja, Copacabana (about 10 km S of Mapiri), *Krukoff 11221* (NY, S, U, US).

Loreya gracilis is similar to *L. umbellata* in leaf shape and indument and branching pattern of the inflorescence. It differs from this and all other *Loreya* species in its much smaller flowers and the color change of the stamens from yellow to red (*Nee & Solomon 30308*). This change in color occurs frequently in *Miconia,* but is not found elsewhere in the *Loreya* alliance (*Bellucia, Henriettea, Henriettella, Myriaspora*).

7. Loreya subrotundifolia (Wurdack) Renner, comb. nov. Figs. 13E, 14A–C, 19.

Bellucia subrotundifolia Wurdack, Phytologia **45:** 331. 1980b. Type: French Guiana. Saül, 3°37′N, 53°12′W, *de Jong 2* (holotype: WAG; isotype: US!).

Small *tree,* 3–5–15 m high. Branchlets sulcate, minutely puberulous, glabrescent; bark gray to brown, smooth. *Leaves* 15–26 × 13–23 cm, orbicular, obtuse or rounded, rounded to truncate or shortly cordate at base, membranaceous, adult leaves glabrous, 5(–7)-nerved, the lateral primaries arising at the blade base; petioles (2–)3–4(–6) cm long. *Cymes* 2–3-branched, 3.5–4 cm long, the peduncle 5–10 mm long. *Flowers* with pedicels 5–10 mm long; hypanthium 5–7 mm long, sparsely puberulous, glabrescent; petals pink, 13–15 × 9–12 mm, ovate or elliptic, apex obtuse; filaments 3.5–5 mm long; anthers 4–5 mm long, 2-pored; style 13–15 mm long; stigma ca. 2 mm diam.; ovary 5(–6)-locular. *Fruit* a glabrous, immature purple, mature dark blue berry, ca. 1.2 cm diam. *Seeds* wedge-shaped with an elongate lateral hilum, bulging testa cells forming an irregular pattern.

Distribution and ecology (Fig. 17): *Loreya subrotundifolia* grows along savanna margins or in open secondary vegetation; it occurs in central French Guiana and in northeastern Brazil.

Specimens examined. FRENCH GUIANA. Saül, *Granville 2815* (CAY, P), *4489* (P, US), *4629* (CAY, P, U, US).

BRAZIL. AMAPÁ: Mun. Mazagão, BR 156, rd. under construction which will connect Macapá with Monte Dourado, 79 km WSW of Macapá, ca. 10 km SW of Rio Preto, *Rabelo et al. 3048* (MG, US).

Although it was noted in the original description of this species that it had a "bud calyx much shorter than typical of the genus and the stigma smaller," nevertheless it was placed in *Bellucia.* While the calyx morphology would certainly not be a sufficient argument for transferring the species, several other features necessitate a transfer: the 5-lobed stigma is unlike that in any other species of *Bellucia*; the epidermal cuticle of the lower leaf surface is smooth instead of papillose as in all species of *Bellucia,* the calyx consists of a truncate limb with vestigial lobes, as is the rule in *Loreya,* rather than well developed sepals or a calyptra as in typical *Bellucia,* and the fruits are purple rather than greenish yellow (as in all species of *Bellucia*). Furthermore, the seeds (Fig. 13E) are wedge-shaped and have large concave fields on the surface as in most other *Loreya* species, but unlike *Bellucia* seeds which do not have such concave areas on the surface. The architecture of the trees of *L. subrotundifolia* conforms to the Aubréville model in the classification of Hallé et al. (1978; cf. VEGETATIVE STRUCTURES). The same growth type is found in *Loreya mespiloides,* but not in any of the arborescent species of *Bellucia* (pers. obs.). Finally, the petals are pink on both sides, a feature not found in *Bellucia,* but present in *Loreya.*

8. Loreya riparia Renner, sp. nov. Type: Brazil. Amazonas: Manaus, rd. towards Praia Dou-

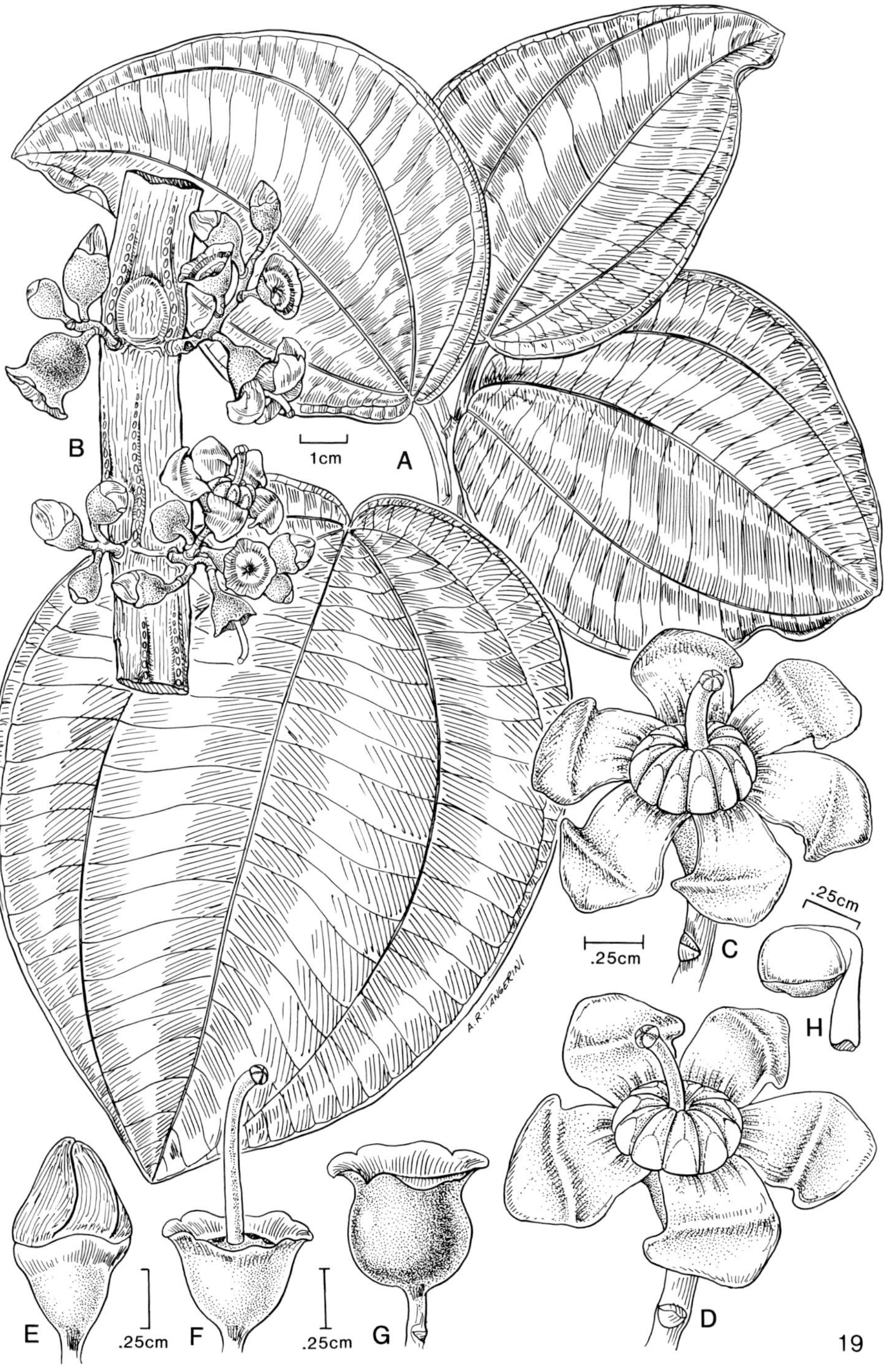

FIG. 19. *Loreya subrotundifolia.* **A.** Habit. **B.** Inflorescences. **C, D.** Flowers. **E.** Bud. **F.** Flower just after anthesis. **G.** Fruit. **H.** Stamen.

rada, right hand side, nr. streamlet, 15 Nov 1984 (fl), *Renner 1015* (holotype: INPA!; isotypes: HBG!, US!). Figs. 12, 14D–F.

Loreyae subrotundifoliae (Wurdack) Renner affinis, foliis inferne strigulosis petiolis longioribus floribus majoribus fructibus flavis differt.

Tree, 8–25 m high. Branchlets very slightly to prominently sulcate, with a conspicuous brown hair-covering; bark gray, smooth. *Leaves* 15–26 × 13–23 cm, suborbiculate, acuminate, cordate at base, membranaceous, mid-vein above strigulose in young leaves, adult leaves glabrous above, strigulose below, 5-nerved, the lateral primaries arising at the blade base; petioles 8–10 cm long. *Cymes* 2–3-branched, 4–5 cm long, the peduncle short, 5–10 mm long. *Flowers* with pedicels 3–7 mm long; hypanthium 9–10 × 11–14 mm, densely puberulous; petals white suffused with pink towards the apex, 15–19 × 10–13 mm, ovate or elliptic, apex rounded or irregular; filaments 4–5 mm long; anthers 6–7 mm long, 2-pored; style ca. 15 mm long, stigma 2–3 mm diam.; ovary 5(–6)-locular. *Fruit* a puberulous, glabrescent, yellow berry, 1.2–1.5 cm diam. *Seeds* wedge-shaped with an elongate lateral hilum, bulging testa cells forming an irregular pattern.

Distribution and ecology (Fig. 17): *Loreya riparia* is endemic in the Manaus area (Brazilian Amazonia) where it occurs along black water creeks, often with the roots inundated.

Paratypes. From the same tree as the type, *Nelson & Nelson 1444* (INPA); Manaus–Caracaraí rd., km 64, ZF-3 rd., km 34; area of INPA-WWF project "Minimal critical size of ecosystems"; along igarapé running through camp "34," *Renner 930* (INPA).

Loreya riparia flowers in late November/December, at the onset of the rainy season, and fruits from the end of December until February. New leaves are put out at the end of the dry season in August/September. The flowers are visited and pollinated by large bees, e.g., *Xylocopa frontalis* (Olivier, 1789).

In overall aspect, the species is similar to *Loreya subrotundifolia* (Fig. 19); it differs in the larger flowers, the denser pubescence of the branchlets, inflorescence, and lower leaf surface, and the yellow fruits (vs. purple fruits in the former species).

9. Loreya strigosa Gleason, Bull. Torrey Bot. Club **60:** 386. 1933. Type: Brazil. Mato Grosso: Plateau between rivers Machado and Madeira, very old clearing, *Krukoff 1593* (holotype: NY!; isotypes: F!, G!, MO!, P!, S!, U!, US!).

Loreya quadrifolia Gleason, Bull. Torrey Bot. Club **60:** 386. 1933. Type: Brazil. Mato Grosso: Nr. Tabajara, upper Machado river region, *Krukoff 1510* (holotype: NY!; isotypes: COL!, G!, GH!, MO!, P!, S!, U!).

Small *tree,* 4–10 m high. Branchlets and branches terete, branchlets densely villose, hairs to 5 mm long. *Leaves* 27–36 × 9.5–18 cm, narrowly elliptic, apically sometimes dentate, acuminate to cuspidate, acute at base, membranaceous, adult leaves moderately to densely sericeous on both surfaces and on the veins, 5-plinerved, the inner pair of lateral primaries departing 5–7 cm above the blade base; petioles 0.5–0.8 cm long. *Flowers* in sessile clusters; pedicels 5–10 mm long; hypanthium 5–6 mm long, densely strigose with basally broadened hairs, 2–4 mm long; petals white, 10–12 mm long, 4–4.5 mm wide at the base, broadly triangular, subacute, bearing on the back a stripe of strigose hairs; filaments 6–7 mm long; anthers 5–6 mm long, 2-pored; style ca. 19 mm long; stigma 1.7–2 mm diam.; ovary 5-locular. *Fruit* a loosely strigose, immature red, mature dark blue berry, ca. 1 cm diam. *Seeds* wedge-shaped with an elongate lateral hilum, bulging testa cells forming an irregular pattern.

Distribution and ecology (Fig. 20): *Loreya strigosa* occurs along the southwestern tributaries of the Amazon from Colombia to the states of Acre, Rondônia, Mato Grosso, and Amazonas in Brazil. The species is known from primary forest at low elevations to 450 m.

Specimens examined. BRAZIL. ACRE: Km 22 from Rio Branco on Rio Branco–Quixada rd., *Lowrie et al. 587* (GH, INPA, NY, US); Porangaba, Rio Juruá-Mirím, *Prance et al. 13168* (F, GH, INPA, M, MG, NY, US). AMAZONAS: Rio Cunhuá at Deni Indian village, 6°43′S, 66°47′W, *Prance et al. 16494* (F, GH, INPA, M, MG, NY, U, US). RONDÔNIA: Rd. to São Lorenço mines, N bank of Rio Madeira, 10 km above Mutúmparaná, *Prance et al. 8868* (F, GH, INPA, NY, P, U, US).

COLOMBIA. CAQUETA: Florencia, cerro de La Sardina, *Cuatrecasas 8917* (US); Florencia, *Pérez-Arbeláez 753* (COL, US).

The type of *Loreya quadrifolia* is identical to that of *L. strigosa* in all respects, except for the longer hairs on the hypanthium. Absent any oth-

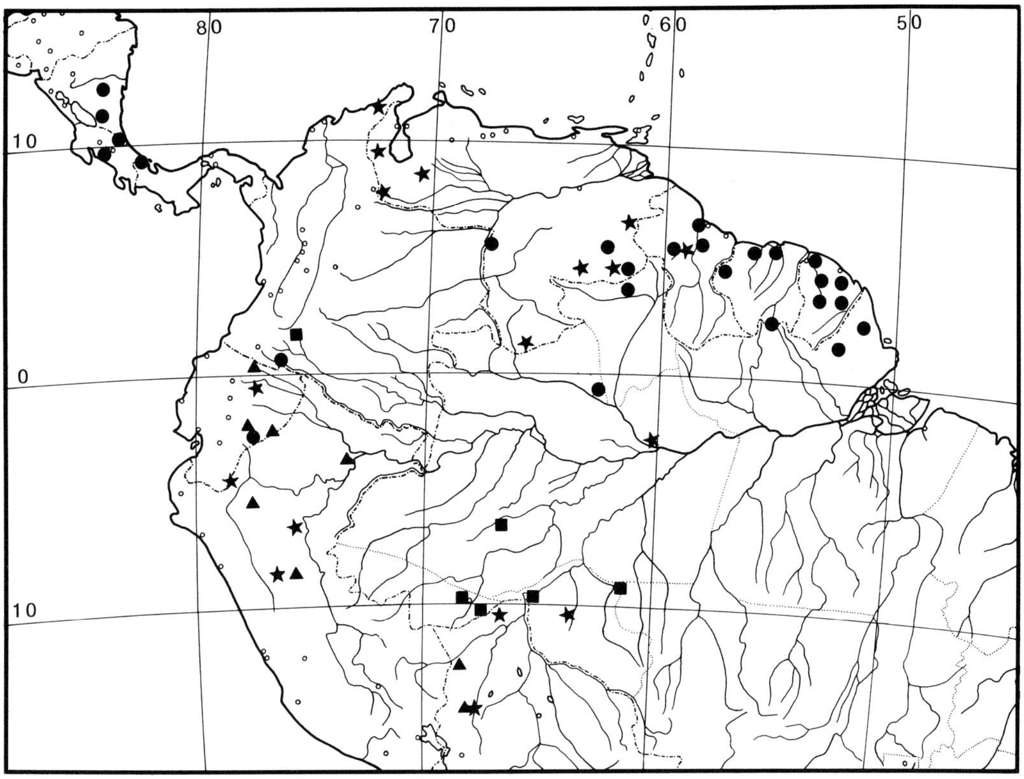

FIG. 20. Distribution of *Loreya*. *L. strigosa* (■), *L. subandina* (▲), *L. mespiloides* (●), *L. spruceana* (★).

er differences, this feature cannot serve to characterize a species in this genus.

10. Loreya subandina Wurdack, Phytologia **18:** 161. 1969. Type: Bolivia. La Paz: Nr. Sarampiuni, San Carlos, Mapiri, *Buchtien 1034* (holotype: US!; isotype: HBG!).

Tree, 6–15 m high. Branchlets angular, strigose, glabrescent; bark light tan, longitudinally fissured. *Leaves* 17–30 × 7–17 cm, elliptic, sometimes slightly dentate, narrowly acute, broadly acute at base, membranaceous, on both surfaces sparsely strigulose, above glabrescent except on the mid-vein and lateral primaries, 7-pli-nerved, the inner pair of lateral primaries departing 4–8.5 cm above the blade base; petioles 2–5 cm long. *Flowers* in subsessile clusters, the peduncle 1–3 mm long; pedicels 10–15 mm long; hypanthium 7–8 mm long, densely strigulose with longer and minute hairs; petals white or pink, 15–18 × 7–9 mm, elliptic, apex irregular; filaments 6–7 mm long; anthers 5–5.5 mm long, 2-pored; style 13–15 mm long; stigma 2.5–3.0 mm diam.; ovary 10-locular. *Fruit* a strigulose, subglabrescent, yellow or red tinged berry, ca. 1.3 cm diam. *Seeds* wedge-shaped with an elongate lateral hilum, bulging testa cells forming an irregular pattern.

Distribution and ecology (Fig. 20): *Loreya subandina* occurs in riparian vegetation, sometimes on periodically inundated soils, in Ecuador, Peru, and Bolivia; low to 700 m elevation.

Representative specimens examined. ECUADOR. MORONA-SANTIAGO: Puerto Morona, Río Morona, 2°52′S, 77°41′W, *Little 584* (LOJA, US). NAPO: Río Napo between Coca (Puerto Francisco de Orellana) and Armenia Vieja, *Harling & Andersson 11968* (AAU, US); Limoncocha, *Zarucchi 2358* (COL, GH, QCA, S, US); 4 km N of Río Napo at mouth of Río Tena, between Tena and Archidona, *Asplund 9181* (B, NY, P, S, US). PASTAZA: Río Bobonaza, Pinacocha, below Montalvo, ca. 76°45′W, 2°17′S, *Øllgaard et al. 34595* (AAU, US).

PERU. AMAZONAS: Quebrada Shimpunts, Río Cenepa, *Ancuash 16* (MO, US). HUANUCO: Aucayacu, *Boeke & Ramirez 1264*(MO, NY, US). LORETO: Iquitos, *Killip & Smith 27455* (US). MADRE DE DIOS: Prov. Manu,

Manu Park, Cocha Cashu uplands, 11°45′S, 71°01′W, *Núnez 6152* (MO, US). PASCO: Oxapampa prov., Palcazu valley, Río San José in the Río Chuchurras drainage, 10°09′S, 75°20′W, *D. Smith 4012* (MO, US).

BOLIVIA. LA PAZ: San Buena Ventura, *R. S. Williams 662* (US).

Local names: sau-chinchak, tunkía (Peru).

Loreya subandina is illustrated in Wurdack (1980a: 319, fig. 33). The species has a 10-celled ovary; in this character and general facies it resembles *L. spruceana*. It differs from that species in its 2-pored anthers and its leaf venation, with the inner pair of lateral primaries diverging 4–8.5 cm above the base (vs. 1.7–2 cm in *L. spruceana*).

11. Loreya mespiloides Miquel, Linnaea **18:** 619. 1844; *Bellucia mespiloides* (Miquel) Macbride, Field Mus. Nat. Hist., Bot. Ser. **13:** 497. 1941; *Loreya brunnescens* (Standley) Gleason, Phytologia **3:** 346. 1950a. Type: Surinam. Para, Vierkinderen, *Focke 946* (holotype: U!).

Henriettea brunnescens Standley, Publ. Field Mus. Nat. Hist. Bot. Ser. **4:** 247. 1929. Type: Panama. Bocas Del Toro: Buena Vista Camp on Chiriquí Trail, 375 m alt., *Cooper 594* (holotype: F!; isotype: US!).

Tree, (2–)4–10(–18) m high. Branchlets sulcate, with a conspicuous brown velvety hair-covering, hairs 2 mm long; bark smooth, gray or tan. *Leaves* 15–30(–40) × 10–20 cm, broadly elliptic to ovate, entire or dentate, acute to acuminate, rounded to broadly acute at base, membranaceous, moderately to densely strigose above, sericeous below, above glabrescent except on the veins, 5(–7)-pli-nerved, the inner pair of lateral primary veins departing 1.5 cm above the blade base; petioles 2–6 cm long. *Flowers* in sessile clusters; pedicels 10–15(–20) mm long; hypanthium (5–)7–8 mm long, loosely to densely strigose, with basally slightly broadened hairs; petals white or pink, 10–15 × 8–10 mm, elliptic, apex obtuse; filaments 5–6 mm long, anthers 5–7 mm long, 1-pored; style 6–8 mm long; stigma 2–3 mm diam.; ovary 5-locular. *Fruit* a densely puberulous, yellow berry, ca. 1.3–2 cm diam. *Seeds* ovoid, the bulging testa cells forming a regular tuberculate surface pattern.

Distribution and ecology (Fig. 20): *Loreya mespiloides* occurs in primary and secondary vegetation on non-inundated soil in Colombia, Ecuador, and throughout southern Central America; it also reaches the Guianas and northeastern Brazil; lowlands to 1100 m elevation.

Representative specimens examined. NICARAGUA. RIO SAN JUAN: Sabalo, 3 km del Río San Juan y del Río Sabalo, *Araquistian 3221* (CAS). ZELAYA: Nueva Guinea, sector Río Demetrio, *Araquistian 3044* (CAS, MO); 5–10 km E of San Antonio on trail to and nearly reaching San Miguel, *Nee & Vega 27856* (CAS).

COSTA RICA. HEREDIA: 2 km E of La Vírgen de Sarapiqui, *Lent 1517* (MO, US). SAN JOSE: Río Negro, Sta. Rosa de Purisal, falda Cerro la Grangecha, *Chacon 1950* (CAS).

PANAMA. Cerro Azul nr. Goofy Lake, *Dwyer 4366* (US).

COLOMBIA. PUTUMAYO: Río San Miguel, Sta. Rosa and vic., *Schultes 3629* (COL).

VENEZUELA. AMAZONAS: ESE of Puerto Ayacucho, 10–30 km on rd. to Gavilan, 5°35′N, 67°22′W, *Steyermark et al. 113894* (US, VEN). BOLIVAR: Icabarú–Sta. Elena rd., *Croat 54198* (NY); Sta. Elena, *Lasser 1648* (F, NY).

ECUADOR. MORONA-SANTIAGO: Puerto Morona, en Río Morona, 2°52′S, 77°41′W, *Little s.n.*, 30 Sep 1975 (COL).

GUYANA. POTARO: Kaieteur Falls, *De La Cruz 4426* (F, GH, MO, NY, US); Pomeroon Distr., Moruka river, *de la Cruz 4565* (F, GH, MO, NY, US); Mazaruni Distr., Rockstone, *Gleason 691* (NY); along Utschi river, between Utschi Falls and mouth, *Tillett & Tillett 45763* (NY, US).

SURINAM. SURINAME: S of Paramaribo, Watramiri, *B.W. 2510,* tree *1578* (GH, NY, U, US). NICKERIE: Fallawatra, *Jiménez-Saa 1504* (US).

FRENCH GUIANA. Piste de Saint-Laurent-du-Maroni, vers Paul Isnard, km 30, *Billiet & Jadin 1714* (BR, US); Massif des Emerillons, face sud, près du sommet de la montagne, *Cremers 6597* (US); Montagne Gros Roche, environs de Boulanger, route de Cacao, *Granville 2999* (AAU, P, U, US); Montagne de la Trinité, sommet NE, *Granville 6312* (B, BR, U, US); Approuague river, Grand Canoti fall, *Oldeman 2755* (P, U, US); Sinnamary, route de Ste. Elie, près de la parcelle ARBOCEL, *Sastre 6017* (U, US).

BRAZIL. AMAPÁ: Mun. Calçoene, BR 156, 96 km NW of Calçoene, ca. 2°48′N, 51°23′W, *Rabelo & Cardoso 2910* (F, MG, NY, US); Perimetral Norte, Mungúba, *Santos 225* (F, MG, NY, US). AMAZONAS: Basin of Rio Demini, vic. of Tototobi, *Prance et al. 10275* (F, GH, INPA, MG, NY, R, U, US). PARÁ: Parque Indigena do Tumucumaque, Rio Parú de Oeste, Missão Tiriyo, 2°20′N, 55°45′W, *Cavalcante 2527* (MG, NY, US). RORAIMA: Serra Tepequem, *Maguire & Maguire 40176* (IAN, NY, US).

The habit of *Loreya mespiloides* is illustrated in Cogniaux (1888, pl. 111), the growth form in Cremers (1983, fig. 9C), and the fruit in Roosmalen (1985, pl. 107). The fruits are dispersed by marsupials and birds (Atramentowicz, 1982; Charles-Dominique, 1986).

Loreya mespiloides is similar to *L. spruceana,* but differs in the 5-celled ovary (vs. the 10-celled one of *L. spruceana*) and the more abundant vegetative pubescence, particularly of the lower leaf surface.

The collection *Pires et al. 50299* (Serra do Navió, Rio Amaparí; Brazil) has much smaller flowers and longer pedicels.

12. Loreya spruceana Bentham ex Triana, Trans. Linn. Soc. London **28:** 143. 1871; *Bellucia spruceana* (Triana) Macbride, Publ. Field Mus. Nat. Hist., Bot. Ser. **13(4):** 499. 1941. Type: Brazil. Amazonas: Nr. Manaus, *Spruce 1249* (holotype: K!; isotypes: BR frag.!; C!, G!, G-BOISS!, GH!, P!; B destroyed, but represented by a photo, F neg. 17253!). Figs. 13A, C; 14G–I.

Loreya collatata Wurdack, Phytologia **18:** 162. 1969. Type: Peru. San Martín: On ridge east of Tingo María, elev. 625–1100 m, *Allard 22575* (holotype: US!).

Small *tree,* 3–10 m high. Branchlets angular, strigose with brownish hairs, glabrescent; bark tan or gray, smooth. *Leaves* 15–26 × (7–)11–20 cm, ovate to broadly ovate, entire or dentate, shortly acuminate, broadly acute to obtuse at base, membranaceous, glabrous on the surface above, puberulous to laxly strigulose beneath, the mid-vein and lateral veins loosely strigulose, 5(–7)-pli-nerved, the inner pair of lateral primaries departing 1.7–2 cm above the blade base; petioles 1.5–4 cm long. *Flowers* in sessile clusters; pedicels 8–10 mm long; hypanthium (6–)7–10 mm long, densely strigulose with minute hairs; petals white, 10–15 × 5–6 mm; filaments 5–6 mm long; anthers 5–6 mm long, 1-pored; style 12–15 mm long; stigma ca. 2.5 mm diam.; ovary 10-locular. *Fruit* a puberulous, basally strigose, yellow berry, ca. 1.5 cm diam. *Seeds* ovoid, the bulging testa cells forming a regular tuberculate surface pattern.

Distribution and ecology (Fig. 20): *Loreya spruceana* occurs from Guyana in a semicircle around the Amazon basin to Bolivia and also in central Amazonia (near Manaus). It grows in primary forest on well-drained upland soil and in more open forest on white sand soil or waterlogged soil, also in secondary vegetation; lowlands to 700 m elevation.

Representative specimens examined. COLOMBIA. GUAJIRA: Sierra Nevada de Sta. Marta, Sierra de San Antonio, Bosque de La Cueva, *Cuadros & Gentry 2979* (COL, MO, US).

VENEZUELA. AMAZONAS: Along Río Marawinuma, E of Base Camp, 0°50′N, 66°09′W, *Thomas 3328* (MO, NY, US). BARINAS: 77 km from Bariñas along rd. to San Cristóbal, *Breteler 4046* (G, MO, NY, P, U, US). BOLIVAR: Río Chikanán, ca. 6°13′N, 61°53′W, *Baumrucker & Calzadilla s.n.* (US); El Dorado–Sta. Elena, 30 km from El Dorado, *Bernardi 2104* (G, NY); NE of Serranía Pia-soi, 3°53′N, 62°46′W, *Steyermark 90693* (US); Sierra Ichun, cercanias del Salto María Espuma (Salto Ichun) del Río Ichun, base de la Sierra de Ichun, trib. of Río Paragua, 4°46′N, 63°18′W, *Steyermark 90367* (F, US). TACHIRA: W of El Pinal, W of bridge over Río Frio, *Steyermark 96690* (US). ZULIA: Sierra de Perija, vic. of Quebrada Koshida, S of Misión de Los Angeles de Tokuku, SW of Machiques, *Steyermark 99940* (B, M, US).

GUYANA. MAZARUNI DISTR.: Garraway Stream, Potaro river, *Maguire & Fanshawe 22967* (GH, U, US).

ECUADOR. NAPO: Confl. of Quiwado and Tiwaeno rivers, *Davis & Yost 937* (F, GH, US); San José de Payamino, 0°30′S, 77°20′W, 40 km W of Coca, *Irvine 128* (F, US). MORONA-SANTIAGO: Taisha, *Cazalet & Pennington 7724* (NY, US).

PERU. AMAZONAS: Río Cenepa, vic. of Huampami, ca. 5 km E of Chavez Valdivia, ca. 78°30′W, 4°30′S, *Ancuash 1123* (F, MO, US); Quebrada Nahem, *Kayap 1072* (MO, US). HUANUCO: Distr. Rupa Rupa, al E de Tingo María, *Schunke V. 10093* (NY, U, US). LORETO: Maynas, trail from Indiana on Río Amazonas to Río Napo, *Gentry et al. 22194* (F, MO, US); Iquitos, *Killip & Smith 27455* (NY). SAN MARTIN: Dtto. Tocache Nuevo, Fundo del Sr. Luis Ludena, *Schunke V. 8111* (US).

BRAZIL. AMAZONAS: 3rd Neblina Exp., Rio Cauaburí, Rio Maturacá, between Camp II and maloca, *N. T. Silva & Brazão 60742* (MG, NY); Manaus, *Ducke 1464* (GH, US), *Prance et al. 4699* (F, GH, INPA, MG, U, US), *Renner 131* (HBG), *237* (HBG), *597* (US), *963* (INPA). RONDÔNIA: Mun. Ariquêmes, 10°35′S, 63°35′W, SW de Ariquêmes, *Teixeira et al. 365* (INPA, MG, NY, INPA).

BOLIVIA. BENI: Rurrenabaque, *Cardenas 1199* (NY). LA PAZ: San Buena Ventura, *R. S. Williams 662* (NY). PANDO: Nicolás Suárez, junto a la barraca San José, *Fernández Casas & Susanna 8255* (MO, NY).

Local names and uses: nispero blanco, misperillo, yujash (Peru); yuhach, yujash, and quisa muyu aula yura (Ecuador), which latter means "small clay pot grandmother tree" (*Irvine 128, 408*). On another collection from Ecuador (*Davis & Yost 937*) the label says that the species is important food for tapirs and wild pigs and is called "titemeneka" meaning "the tapirs wild cacao."

In central Amazonia (near Manaus) and in Venezuelan Amazonia, I have observed several species of bees [*Xylocopa frontalis* (Olivier, 1789), smaller species of *Xylocopa* and *Bombus, Euglossa* spp., *Eulaema moczaryi* (Friese, 1899), and *Eulaema* cf. *meriana*] pollinating the flowers of this species (Renner, 1984a, in press).

Loreya spruceana shows considerable variation in the length of the calyx limb and the length of the hairs, particularly on the petiole; the calyx in central Amazonian specimens is 5–6 mm long, that of extra-Amazonian ones is mostly 1.8–2 mm long. However, these characters do not covary with other traits. *Loreya spruceana* is illustrated in Wurdack (1973: 660, fig. 58, as *L. collatata*).

13. Loreya klugii Renner, sp. nov. Type: Colombia. Putumayo: Umbria, 00°54′N, 76°10′W, 325 m alt., Jan–Feb 1931 (fl), *Klug 1927* (holotype: F!; isotypes: BM!, GH!, MICH!, MO!, NY!, S!, US!). Figs. 13B, 21.

Loreyae spruceanae Bentham ex Triana affinis, foliis minoribus floribus minoribus ovariis quinquelocularibus differt.

Tree, 3–8 m high. Branchlets slightly angular or sulcate, branches terete, bark pale tan, smooth. *Leaves* 17–19 × 6.5–9 cm, elliptic, acuminate, acute at base, membranaceous, intercostal areas glabrous above and beneath, mid-vein and lateral primaries strigose on both surfaces, 5-plinerved, the inner pair of primaries diverging 1.5–3 cm above the blade base; petioles 1.5–3 cm long. *Flowers* in sessile clusters; pedicels 5–7 mm long; hypanthium ca. 4 × 4 mm, strigulose; petals white, 6–7 × 3–4 mm, elliptic, apex obtuse: filaments 4–5 mm long; anthers 3–4 mm long, 1-pored; style 11–12 mm long; stigma capitate, ca. 2 mm diam.; ovary 5-locular. *Fruit* a puberulous, greenish yellow berry, ca. 1.2 cm diam. *Seeds* ovoid, the bulging testa cells forming a regular tuberculate surface pattern.

Distribution and ecology (Fig. 17): *Loreya klugii* occurs in subandean forest in Colombia and Peru at low elevations to 400 m; it has been collected in primary and secondary vegetation.

Additional specimens examined. COLOMBIA. PUTUMAYO: Topotypical, *Klug 1968* (BM, F, GH, MICH, NY, US).

PERU. AMAZONAS: Quebrada Chichijam Entsa, Río Cenepa, *Ancuash 434* (MO, US). LORETO: Río Yuvineto, affl. Río Putumayo, territoire des indiens Secoya, Sutukapowehe purma, *Barrier 585* (G, P, US); Maynas, Río Yuvineto, affl. Río Putumayo, territoire des indiens Secoya, *Haxaire 937* (G, P, US); Dtto. Indiana, Quebrada Yanayacu above Bombonaji, *McDaniel 17210* (US).

BOLIVIA. COCHABAMBA: Prov. Carrasco, 62 km de Villa Tunari hacia Puerto Villarroel "Parque Litoral" cerca de Ivirgarzama, *Beck 1488* (US).

Local names: rifarillo (Peru).

The species is named after the collector of the type, Guillermo Klug (?–1946), who also collected the type of *Loreya umbellata.* Some sheets of the type collection had already been annotated by H. A. Gleason as "*Loreya* sp. nov.?" *Loreya klugii* is a typical member of the genus in the relatively large, membranaceous, dark brown leaves as well as in the flowers.

Excluded Taxa

Loreya grandiflora O. Berg ex Triana, Trans. Linn. Soc. London **28:** 143. 1871, nomen nudum.

Loreya umbelluliflora O. Berg ex Triana, Trans. Linn. Soc. London **28:** 143. 1871, nomen nudum.

Loreya fasciculiflora Naudin, Ann. Sci. Nat. Bot. III, **18:** 110. 1852.
=*Henriettella sessilifolia* (L.) Triana, Trans. Linn. Soc. London **28:** 143. 1871.

Loreya flavescens (Aublet) Naudin, Ann. Sci. Nat. Bot. III, **18:** 110. 1852.
=*Henriettella flavescens* (Aublet) Triana, Trans. Linn. Soc. London **28:** 143. 1871.

Loreya huberi Wurdack, Phytologia **50:** 306. 1982.
=*Bellucia huberi* (Wurdack) Renner, Mem. New York Bot. Gard. **50:** 28. 1988.

This species is better accommodated in *Bellucia* because it has the papillose lower leaf surface typical of that genus; also, in its sometimes axillary 5–6-merous flowers with 8-lobed stigmas (vs. usually 5-merous flowers with 5-lobed stigmas borne on the old wood in *Loreya*) and in its seeds it fits in better with *Bellucia* than with *Loreya.*

Loreya mucronata Gleason, Mem. New York Bot. Gard. **8:** 143. 1953. Type: Venezuela. Amazonas: Base of Cerro Sipapo, *Maguire & Politi 28168* (holotype: NY!; isotypes: F!, MO!, US!).
=*Henriettea mucronata* (Gleason) Renner, comb. nov.

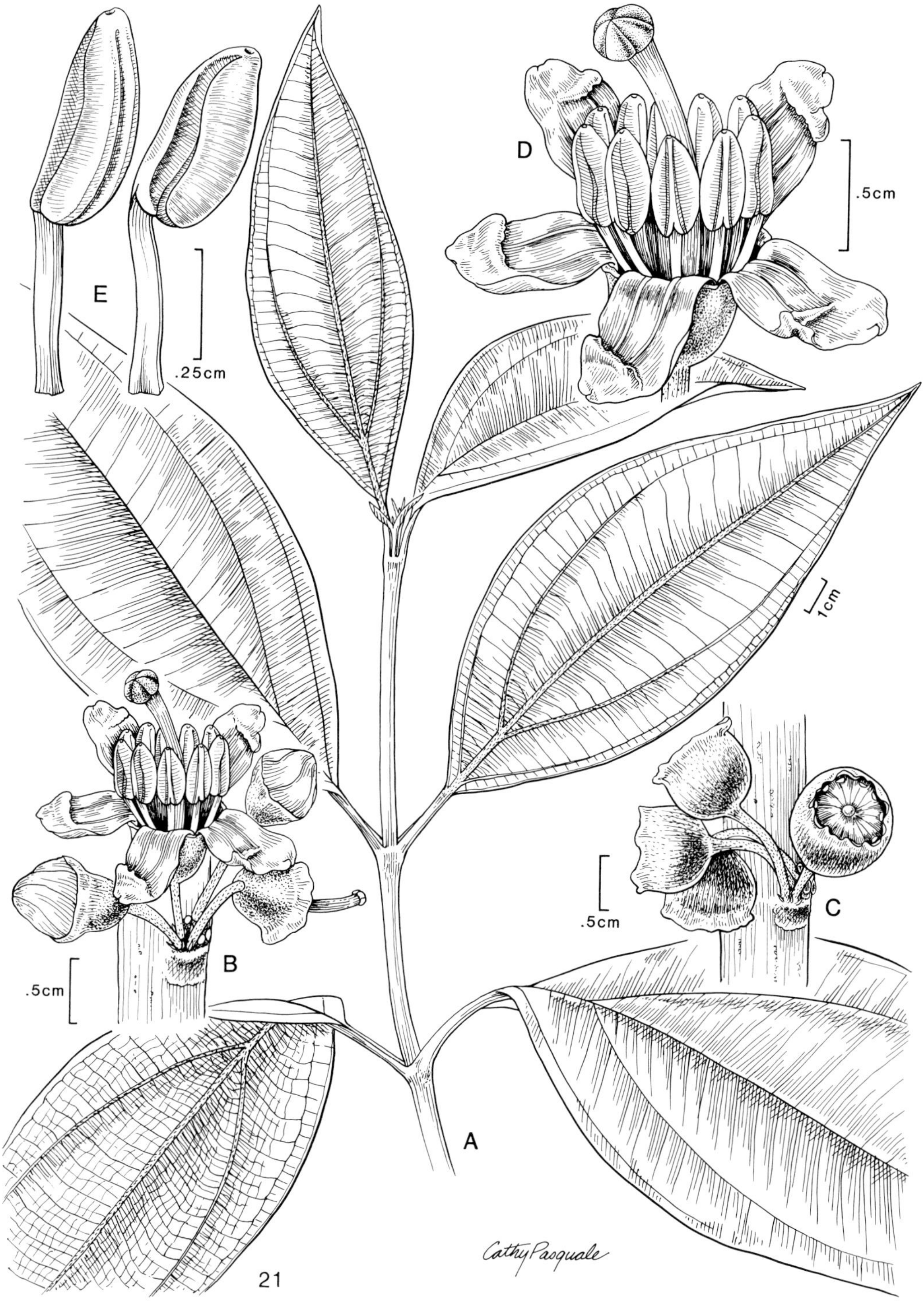

FIG. 21. *Loreya klugii.* **A.** Habit. **B.** Inflorescence. **C.** Fruits, drawn from *Beck 1488.* **D.** Flower, drawn from *Klug 1927.* **E.** Stamens.

The leaves of *Henriettea mucronata* are 3-plinerved (with an additional pair of submarginal veins), the flowers are 6-merous with the connective prolonged dorsi-basally, the fruits are urceolate with short sepals, and the seeds (in SEM) have the pronounced lateral hilum surrounded by a rim and the long, thin, tuberculate testa cell wall projections typical of *Henriettea.*

Loreya trinitensis Crueger, Linnaea **20:** 108. 1847. =*Henriettella sessilifolia* (L.) Triana, Trans. Linn. Soc. London **28:** 143. 1871.

Taxonomic Revision of *Macairea* (Tibouchineae)

Introduction

Macairea is a genus of shrubs and subshrubs numbering 22 species. It is most diverse in lowland savannas and on sandstone mountains in the states of Amazonas and Bolívar in Venezuela. Very few species occur south of the Amazon river; only one is endemic to the Brazilian Planalto, where it is common and widespread. All species grow in open habitats on poor, acidic soils.

The need for a revision of *Macairea* has been apparent for many years (cf. **General Introduction**). It was last treated by A. Cogniaux (1891a), who recognized 14 species. Since then, numerous (25) new species have been described and much collecting has been done in those areas in which the greatest *Macairea* diversity is seen. Thus, the situation has improved since Gleason (1934), treating "one of the best defined and least known genera in the family" for northern South America, wrote that more than half of the 28 species of *Macairea* he recognized were represented only by the type. I have examined about 1000 collections and was able to study a few species in their natural habitat during fieldwork in Brazil and Venezuela. However, there are still two species known only from one or two collections.

Taxonomic History

Bonpland (1820) described the first species of *Macairea* as *Rhexia radula*; a colored plate accompanied the detailed description in his *Rhexiae.* De Candolle, when working on the Melastomataceae for the *Prodromus* (1828a) and his *Mémoire sur la Familles des Mélastomacées* (1828b), studied the material at Paris and established the genus *Macairea* to include Bonpland's species and three new ones, one based on a collection by Alexandre Rodrigues Ferreira and two from material sent to him by Martius. He named the genus in honor of Isaac F. Macaire (1796–18??), a Swiss phytochemist particularly interested in the effects of opium on the senses. De Candolle (1828b) thought the genus "un des plus naturels de tous ceux que j'ai eu occasion d'observer"; it is thus perhaps not surprising that all except one of the 44 entities described in *Macairea* to date are regarded as clearly belonging there.

Bentham (1840) described four new species based on material brought back by Robert and Richard Schomburgk from their travels in Guayana. Two of these collections, deposited in Paris, were redescribed by Naudin (1850a, 1850b). Based on material collected by Spruce in Brazil, Peru, and at the Venezuelan–Colombian border, Triana (1871) named three species. Cogniaux, in his treatment of the family and subsequent publications (1885, 1888, 1896, 1908, 1912), recognized all the earlier species and described ten additional ones, mostly from Brazil. Pilger (1905), Huber (1914), Hoehne (1922), Ducke (1922), and Brade (1958, 1959a, 1959b) named eight more species from Brazil; however, none of these and only one of the species described by Cogniaux merit specific status in my opinion.

In Guayana and Venezuelan Amazonia the picture is different; there the exploration of remote areas has brought to light many new species of *Macairea* and additional species may exist on as yet unvisited mountains (called "tepuís" in the state of Bolívar and "cerros" in Territorio Federal Amazonas). Expeditions by F. V. McConnell and J. J. Quelch to Roraima, by G. H. H. Tate to Mt. Duida, by F. Cardona to the Chimantá Massif, and the extensive collecting program of the New York Botanical Garden carried out by B. Maguire, J. J. Wurdack, G. S. Bunting, and W. M. Keith resulted in the discovery of ten new species of *Macairea.* These were described by Brown (1901), Gleason (1931b), and Wurdack (1958a, 1958b, 1964, 1970). Many additional records of *Macairea* species in Venezuela are due to recent efforts of J. Steyermark, O. Huber, and others.

The selection of a generotype from the four species placed in the genus by de Candolle (1828a, 1828b) presents some difficulty. Gleason (1934) thought that de Candolle must have had *Macairea thyrsiflora* or *M. rufescens* in mind because the generic description in Gleason's words "calls for glabrous leaves" on the surface. In fact, de Candolle says (1828a) "Folia mucronata coriacea integerrima supra glabra." De Candolle (1828b) discusses and illustrates this (pl. 6, fig. 4): "la surface supérieure est tantôt glabre (*M. thyrsiflora, M. rufescens*), tantôt couverte de poils epais, roides, serrés, qui lui donnent l'apparence d'une brosse (*M. radula* et *adenostemon*)." The figure shows the stiff, bristlelike hairs on the leaf surface of *M. radula* seen from the side. In *M. thyrsiflora,* the leaf surface has much shorter and stouter conical hairs appearing as short hard points (i.e., mucronate). De Candolle was clearly aware of the variation in the roughness of the leaf surface among the material he examined: the unfortunate phrasing in the *Prodromus* possibly was a slip of memory. I have selected *Macairea radula* (Bonpland) DC. as the generotype; it is the first species known and has the advantage of being illustrated.

Morphology and Anatomy

Habit, Wood, and Bark

Most species of *Macairea* are long-lived, erect shrubs, 1–2 m tall; *M. chimantensis, M. pachyphylla, M. thyrsiflora, M. rigida,* and *M. stylosa* sometimes become treelets of 5–12 m. Three species (*M. multinervia, M. lasiophylla, M. philipsonii*) are suffrutescent perennials, between 10 cm and 1 m high, and described as "subshrubs" or "herbs" by collectors. Branching is sympodial with lateral, nearly erect, primary branches initiated near the ground. The branchlets are quadrangular. The nodes are thickened and—when hairs are present—bear longer hairs than the rest of the branchlet. Growth is periodic, as indicated by the varying length of internodes; field observations on *M. rufescens, M. thyrsiflora,* and *M. theresiae* showed the greatest flush of vegetative growth to occur after flowering.

Anatomically, the woods of *Macairea axilliflora, M. maroana, M. neblinae, M. pachyphylla,* and *M. thyrsiflora* are typical Tibouchineae wood (Van Vliet et al., 1981; Welle & Koek-Noorman, 1981) with average vessel member lengths of 325–555 μm, fiber/vessel ratios of 1.22–1.50, scarce paratracheal parenchyma bands, and uni- or biseriate rays. These authors found *M. axilliflora* to differ strongly from the other species in the number and diameter of the vessels; this species occurs in savannas along riverbends which are regularly inundated for prolonged periods.

The bark in *Macairea* is gray or brown, rough, longitudinally fissured, and usually exfoliates in thin strips on the older branches. In *Macairea theresiae,* the older branches become twisted (screw driver fashion) and the bark exfoliates in long shreds.

Indumentum and Glands

At least six kinds of multicellular hairs are found in the indumentum of *Macairea* species. First, there are two types of eglandular hairs, differing primarily in stoutness and pliability: some are short, stout, and conical, with broad bases (stiffened with sclerenchyma fibers) and short, often downward-curved tips. They are found on the leaf surface of many species (Fig. 23A). The second type of eglandular hair is more slender and flexuous; it is illustrated by an SEM-micrograph of hairs of *Macairea spruceana* in Wurdack (1986, fig. 76). The slender hairs occur on leaves, stems, and inflorescences and form a hirsute or sericeous covering. They range in color from yellowish to rust and are often apically slightly wavy.

Second, there are two types of glandular trichomes, relatively long ones (1–10 mm) and very short ones. The relatively long ones are stout (Wurdack, 1986, fig. 47, *Macairea parvifolia*) or subulate (Fig. 23D) and are homologous to the eglandular hairs described above, except for terminating in a globose, multicellular, glandular head, which may be caducous. Long glandular hairs are found in 13 of the 22 species of *Macairea* and make the plants feel viscid or even mucilaginous (fide Gleason, 1931b: 298, referring to *M. lanata*). The stalks of the short glandular hairs consist of a few or only two cells and therefore, the globose gland heads appear sessile at low magnification (Figs. 22B, D; 23C); I am referring to them as "sessile glands" in the species

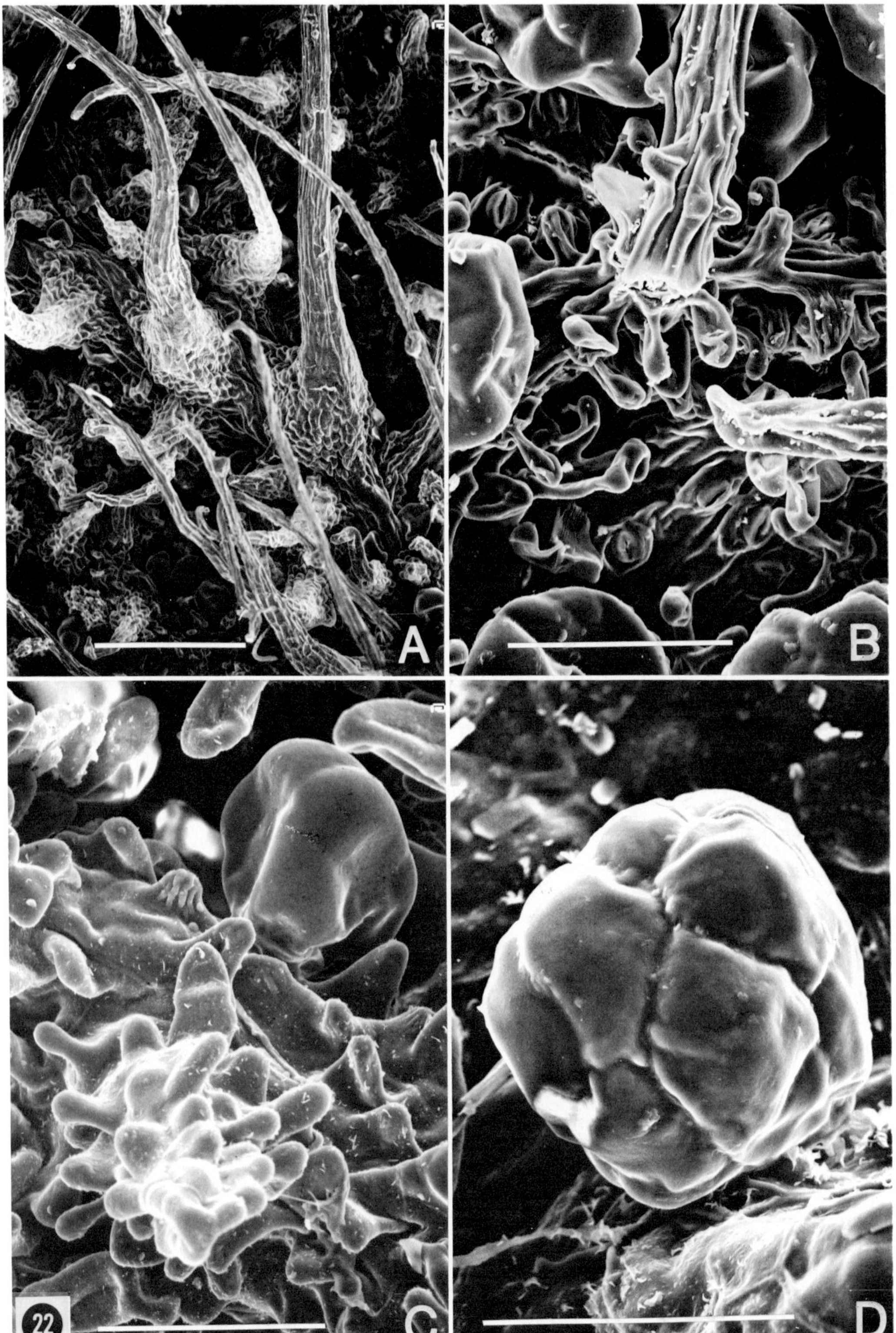
A
B
22
C
D

descriptions. The heads of these short-stalked glands measure 60–80 μm in diam. in *Macairea spruceana, M. rufescens, M. theresiae, M. maroana, M. sulcata, M. thyrsiflora, M. chimantensis,* and *M. duidae.* In *M. axilliflora,* the gland heads are much smaller, 40 μm diam.

Third, there are erect hairs with a well-developed axis and a broad, conspicuous base which is roughened due to a moderate number of short arms with blunt tips (Fig. 22A–C).

Finally, there are short, few-branched, papillae-like hairs with thin walls, which make the lower leaf surface appear velvety. In scanning electron microscopy, wax platelets were found on the surface of these papillae in *Macairea rufescens* and *M. stylosa* (Fig. 23B, D).

On the vegetative parts of the plants, long glandular hairs and short-stalked ("sessile") glands replace each other with either the one or the other present but never both in the same species, with the exception of *Macairea maroana.* In *M. maroana,* both glandular hair types are present: the lower leaf surfaces, the petioles, the branchlets, and the peduncles bear sessile glands and long eglandular hairs; the pedicels, hypanthia, and calyx lobes have gland-headed long hairs, but no sessile glands.

On seedlings of *Macairea rufescens, M. theresiae,* and *M. neblinae* grown in the greenhouse, the sessile glands or long glandular hairs were found in the position and density typical of the species.

Leaves

Phyllotaxy in *Macairea* is opposite and decussate. Like in most members of the family, the leaves are petiolate, estipulate, and predominantly subcoriaceous or coriaceous. The petioles are short, only in two species do they exceed 2 cm. The venation usually consists of one or two pairs of subparallel lateral primary veins arising from a common point at the base of the blade. Leaves are shortly pli-nerved in *M. stylosa.* The outermost pair of primaries is always close to the leaf margin. In *Macairea rufescens* (Fig. 34), *M. sulcata,* and to some extent in *M. spruceana,* only an inconspicuous pair of lateral primary veins is present and the leaves appear almost pinnately veined (they are called "pseudo-pinnately veined"). Three to four pairs of lateral primary veins are found in *M. cardonae* and *M. multinervia.* The primaries are elevated on the lower foliar surfaces and usually impressed on the upper surfaces. In some species, thick, loosely reticulate veinlets are clearly visible on the lower surface.

The leaf margins are entire; only in *Macairea maroana* are they sometimes denticulate. Leaf shape ranges from linear (*M. linearis*) and elliptic to broadly ovate, cordate, or obovate. In *Macairea,* foliar shapes, particularly the shape of the leaf apex and base, are of great systematic value. Leaf pubescence varies from almost none (*M. axilliflora, M. linearis*) through moderately hirsute with glandular or eglandular hairs to sericeous or strigose (see above *Indumentum and Glands*). The abundance and the length of the hairs vary considerably between the proximal and distal and central and marginal sections of an individual leaf, with the age of the leaf, and with its origin. However, the various combinations of the different types of hairs are diagnostic for each species.

Small crystals and druses of calcium oxalate, few in number, have been reported in the leaves of *Macairea radula, M. rufescens,* and *M. spruceana* (Pflaum, 1897: 77–79, for an English summary of Pflaum's results, see Metcalfe & Chalk, 1979).

In his extensive survey of aluminum-accumulating plants, Chenery (1948, and in a letter to J. Wurdack where the names of the three species of *Macairea* analyzed are given) found that *M. thyrsiflora* and *M. lasiophylla,* but apparently not *M. theresiae,* accumulate considerable amounts of aluminum in their leaves.

Inflorescences and Flowers

In most species, the *inflorescences* are terminal paniculate cymes bearing from five to 75 flowers.

←

FIG. 22. Scanning electron micrographs of lower leaf surfaces in *Macairea.* **A, B.** *M. thyrsiflora* (*Maguire et al. 41800*; A bar = 200 μm, B bar = 50 μm). **C.** *M. thyrsiflora* (*Maguire et al. 53663*; bar = 50 μm). **D.** *M. chimantensis* (*Wurdack 34177*; bar = 40 μm).

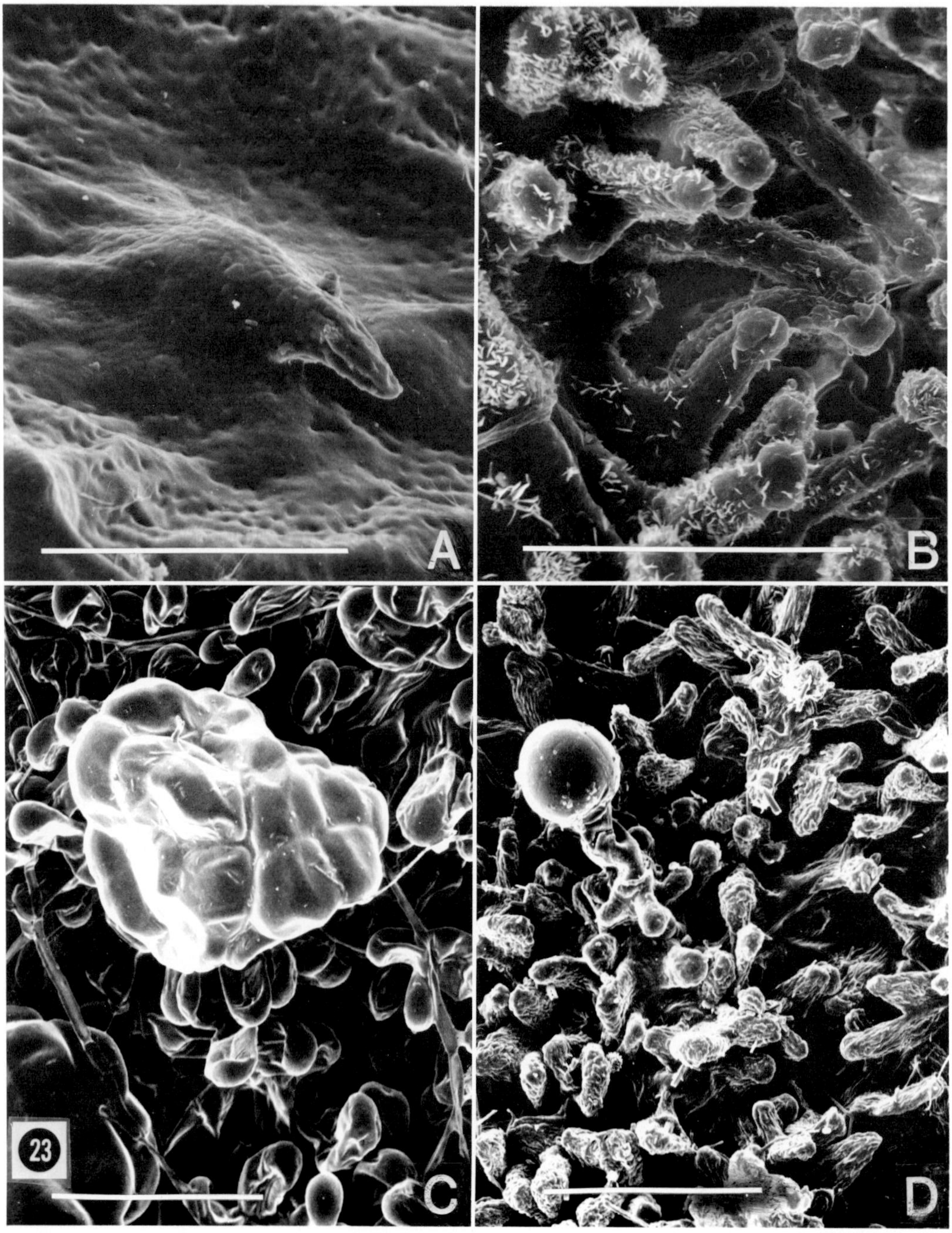

FIG. 23. Scanning electron micrographs of leaf surfaces in *Macairea*. **A.** *M. thyrsiflora,* upper surface (*Maguire & Wurdack 34503*; bar = 200 μm). **B.** *M. rufescens,* lower surface (*Maguire et al. 41709*; bar = 50 μm). **C.** *M. theresiae,* lower surface (*Rodrigues & Coêlho 7908*; bar = 50 μm). **D.** *M. stylosa,* lower surface (*L. Williams 14298*; bar = 50 μm).

In *Macairea lanata* and *M. stylosa,* the cymes are few-flowered and the leaves supporting them are foliage leaf-like. In *M. axilliflora,* the cymes are very condensed so that the inflorescences consist of groups of three to five subsessile flowers, subtended by what appear to be the youngest vegetative leaves; the transition from the uppermost foliage leaves to the lowermost bracts is also very gradual in the inflorescences of *M. multinervia, M. lasiophylla,* and *M. philipsonii.*

In general, the pedicels bear several pairs of bracts and bacteoles. These are membranaceous, often hyaline at the margins, broadly ovate or lanceolate and covered with the same indument as the rest of the inflorescence. Since the bracts and bracteoles vary greatly in size and persistence intraspecifically, they are of little taxonomic value.

Macairea flowers are 4-merous, actinomorphic, and borne erect or slightly tilted. The androecium becomes zygomorphic during anthesis because the stamens decline to one side of the flower. The flowers are perigynous, with a well-developed campanulate hypanthium which is free from the ovary, but fully envelops it. The hypanthium is covered externally with long glandular hairs or else with eglandular hairs and globose sessile glands; in *M. linearis* it is covered with appressed, broadly conical hairs, and in *M. stylosa* and *M. axilliflora* it is glabrous. Eight evenly spaced longitudinal ribs are more or less conspicuous; they represent the vascular supply for the sepals, petals, and stamens.

The sepals are free or rarely united for a short distance, broadly or narrowly triangular, spreading during anthesis and persistent. On the exterior side, they bear the same type of pubescence as the hypanthium; they are glabrous within.

The four petals spread horizontally at anthesis. They are glabrous, except for short-stalked glands along the margin (best seen in buds), obovate with a short claw, and very thin and fugacious, rarely lasting longer than two days. In the majority of species, the petals are deep purple, pink, or magenta (=purplish red) with a white basal blotch. Intraspecific variations in petal color are present in *Macairea lasiophylla* and *M. thyrsiflora,* with white-flowered plants occurring occasionally. In the latter species, it seems that white flowers are the rule in Guyana, whereas elsewhere they are pink. A constant feature of all species with basally white petals is a color change of that section from white to red. This color change is reported many times in 16 different species, and I have observed it in the field in *M. theresiae, M. neblinae,* and *M. rufescens.* Besides the petal bases, it affects the bases of the filaments and the style; starting in the late afternoon of the first day of anthesis, it is completed on the morning of the second day (during which the petals fall off). *Macairea duidae, M. philipsonii, M. axilliflora, M. lanata, M. linearis,* and *M. sulcata* have purple flowers lacking white centers or are incompletely known in this respect (see under species descriptions).

The eight stamens are dimorphic, with the degree of dimorphism varying between and within species. The four larger stamens are inserted in front of the sepals, the four smaller ones in front of the petals. Shortly after expansion of the petals, the filaments (which contain starch grains) all bend to one side of the flower opposing the style. This involves a 180° torsion in the filaments furthest from that side. The filaments are terete, white on the first day, rosy on the second, and in most species ventrally bear short-stalked glands (Figs. 34D, 36H). The filament glands are absent in *Macairea axilliflora, M. philipsonii, M. lasiophylla,* and *M. multinervia*; they are sometimes lacking in *M. lanata.* The connectives are always prolonged and expanded dorsi-basally into a small, cordate or collar-like appendage; the appendages of the antesepalous stamens are larger. In *Macairea lasiophylla,* the connective appendage is very small dorsally, but ventrally has an upward turned lobe on either side of the filament, a feature setting this species apart from all others.

Macairea anthers are linear and the two pollen-sacs open by a single apical pore. In most species, the pollen-sacs unite shortly below the pore into a tube forming a spoon-shaped beak; in *Macairea axilliflora, M. lasiophylla,* and *M. chimantensis* the anthers are truncate.

The gynoecium consists of an ovoid ovary, a sigmoid, thin, terete style, and a punctiform stigma. In all species of *Macairea,* the style arises smoothly from the summit of the ovary and bears some glandular hairs on its mid section (Figs. 36G, 34E). In *M. thyrsiflora,* the ovary apex sometimes forms four lobes surrounding the base of the style, an aberrant feature interesting from a phylogenetic point of view (see INTER- AND

INTRAGENERIC RELATIONSHIPS). The ovary apex is either covered with short-stalked glands or is sericeous with long eglandular hairs. It is usually 4-locular; in a few species the number of locules is reduced to three or two. In some of the latter, the number of ovary cells is not completely stable.

Fruits and Seeds

In *Macairea,* the fruit is a dark brown, semi-woody, campanulate capsule, held upright (Figs. 34G, 36F). It is enveloped by the hypanthial wall and dehisces loculicidally in the upper section. The capsules withstand even strong winds and persist on the plants to a year. Except for differences in size, the fruits are homogeneous and of no utility in the taxonomy of the genus.

The seeds are small, 0.5–1.4 mm long, and dark brown. I investigated seeds of all but one species (*Macairea philipsonii,* of which no seeds were available) using scanning electron microscopy. Figures 24 and 25 show the variation in seed shape and seed coat sculpture encountered. The seeds are cochleate or subcochleate, in two species they are polyhedric (Fig. 24A, C). The hilar region, where the radicle emerges upon germination, is circular or subcircular, clearly defined, and positioned subterminally (Figs. 24F, G; 25A, D). The anticlinal walls of the testa cells are undulate (Fig. 24B) and the periclinal walls form shallow tubercles, which often collapse as the seeds get older (Fig. 24D).

The taxonomic usefulness of seed characters in the Melastomataceae has a long tradition (Don, 1823). Cogniaux (1891a) used the cochleate seed shape and the circular hilum to distinguish the Tibouchineae from the Microlicieae (which have more elongate seeds and an oblong hilum). By these criteria, *Macairea* seeds belong in the tibouchinoid group. Recently, the traditional placement of *Macairea* in the Tibouchineae was called into question by Whiffin and Tomb (1972), who judged that *Macairea* seeds were microlicioid. Their definition (l.c.) of "microlicioid" includes a surface criterion in addition to the shape and hilum criteria used traditionally: "The major characteristic of this seed type is the foveolate surface pattern." Tibouchinoid seeds are defined (l.c.) as being "cochleate to somewhat elongate-cochleate and tuberculate, with the tubercula pronounced to relatively shallow. Usually there is some form of minute surface patterning on the tubercula, which may vary from papillate to muriculate." Since in *Macairea* the testa surface sculpture varies considerably during the development of the seeds, the presence or absence of tuberculae or shallow foveolae (which are in fact collapsed tuberculae) is not a reliable indicator of the tribal affinities of *Macairea.* However, in the minute secondary sculpture, the seeds of *Macairea duidae* (Fig. 24H) and *M. cardonae* (Fig. 25C) conform exactly to the description of Tibouchineae seeds given by Whiffin and Tomb (1972).

GALLS

Large (1–4 cm in diam.), roundish, hairy galls are frequently found on the branchlets of *Macairea radula, M. pachyphylla,* and *M. thyrsiflora.* They have attracted the attention of numerous collectors. Whereas the galls in the first three species seem to be of one kind, the two types of galls found on *M. neblinae* look different (leafy not hairy as the above). These latter galls are caused by Lepidoptera larvae (R. J. Gagné, pers. comm.). The hypanthia of *M. stylosa* are often swollen due to infection by another gall-causing insect.

POLLEN MORPHOLOGY

I acetolyzed pollen of all species, except *Macairea philipsonii* and *M. cuieirasii,* according to procedures outlined in Erdtman (1960). Samples for scanning electron microscopy (SEM) were first carbon-coated and then sputter-coated with gold-paladium and examined with a Hitachi S-570. Slides for light microscopy were made with glycerin jelly and sealed with wax. I measured the pollen grains from light microscope slides, with samples consisting of 7–10 grains. Only one pollen preparation per species was made, and thus, the measurements are indicative only of the generic range in pollen grain size. The terms used here to describe pollen morphology are those of Erdtman (1971) and Patel et al. (1985). Light microscope slides of all samples are deposited at the Palynological Laboratory, Department of Botany, Smithsonian Institution.

Pollen grains of all species examined with SEM are essentially similar in their exine sculpture (Figs. 26–29). The grains are shed as monads,

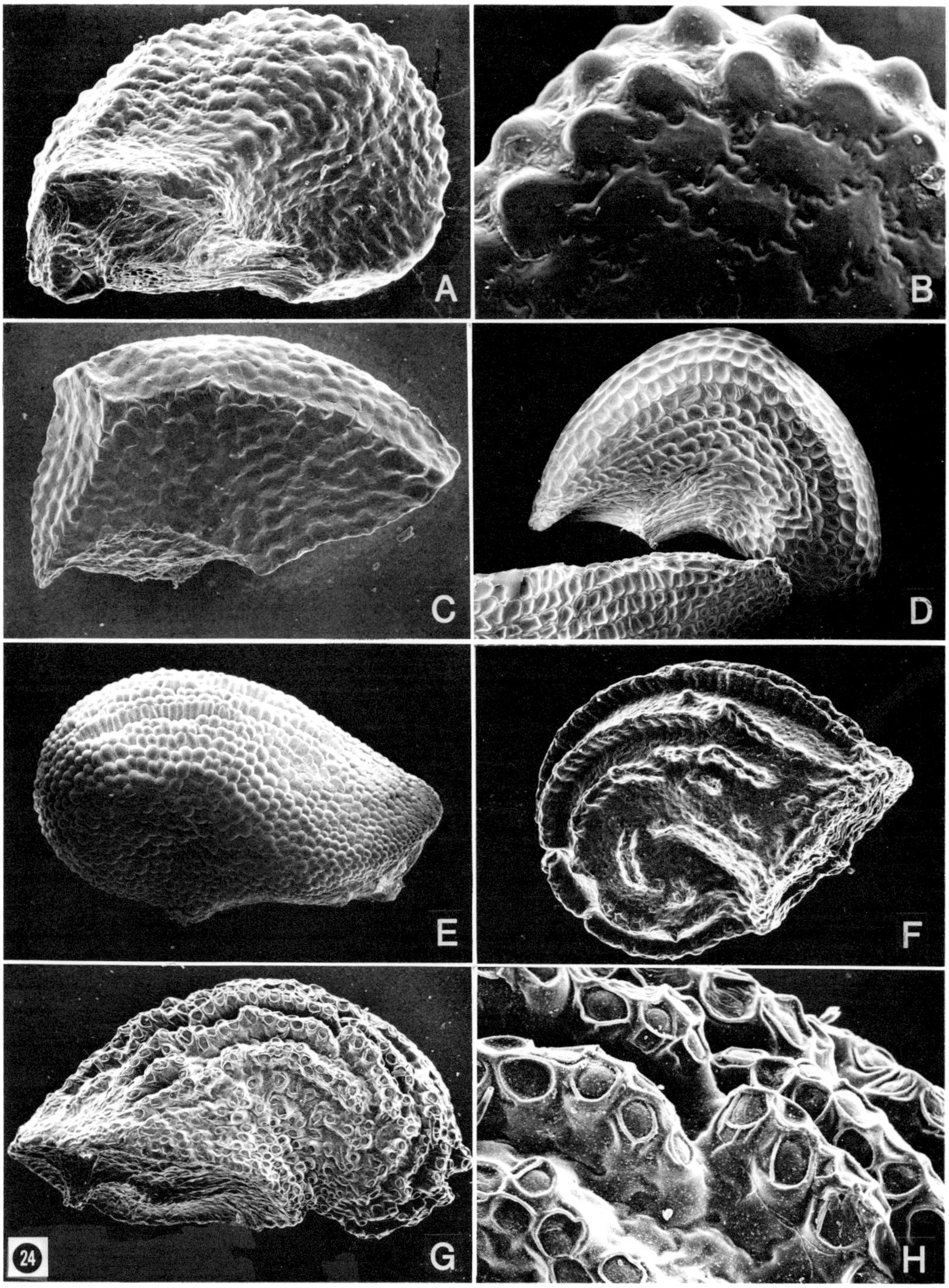

FIG. 24. Scanning electron micrographs of seeds in *Macairea*. **A, B.** *M. axilliflora* (*Huber et al. 3734*; A ×65, B ×250). **C.** *M. stylosa* (*Huber & Medina 5769*; ×50). **D.** *M. rigida* (*Maguire & Maguire 35068*; ×65). **E.** *M. lanata* (*Steyermark et al. 126407*; ×50). **F.** *M. lasiophylla* (*Tate 332*; ×85). **G, H.** *M. duidae* (*Maguire et al. 29644*; G ×60, H ×265).

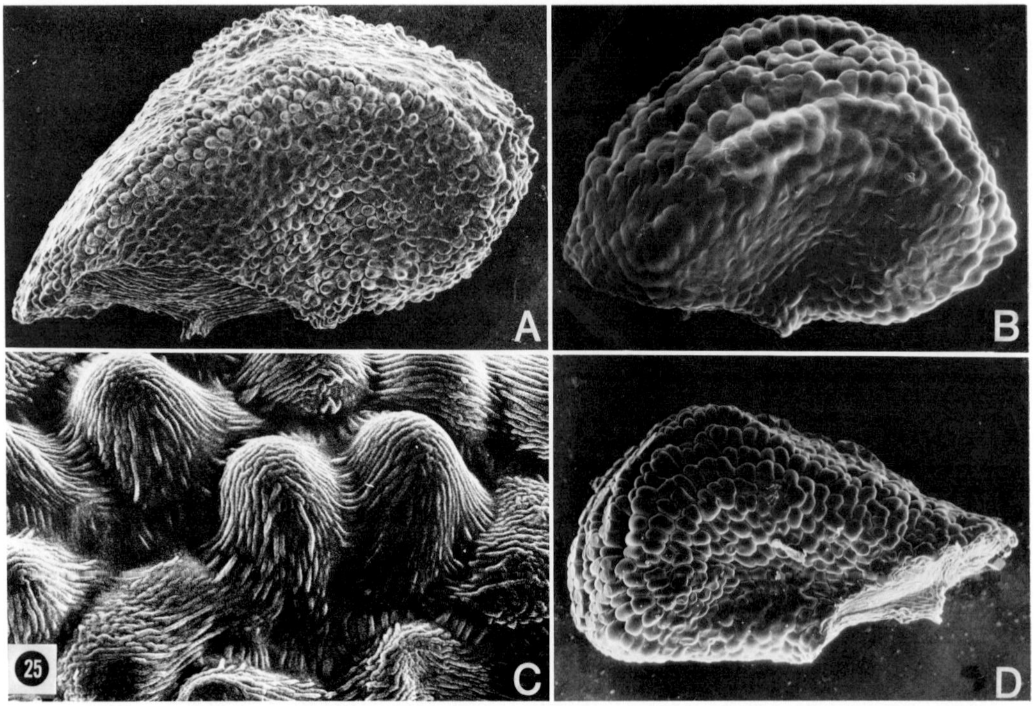

FIG. 25. Scanning electron micrographs of seeds in *Macairea*. **A, C.** *M. cardonae* (*Cardona 256*; A ×50, C ×500). **B.** *M. maroana* (*Pires & Marinho 15709*; ×50). **D.** *M. cuieirasii* (*Rodrigues 6039*; ×50).

radially symmetrical, and isopolar; all have smooth or striate surfaces; and all have 3-colporate apertures alternating with three pseudocolpi ("subsidiary colpi" in the terminology of Patel et al., 1985). The pseudocolpi have different sculpturing from the mesocolpial regions and there are prominent extensions of the mesocolpi over the endoapertures (Figs. 26H, K; 27C, F, I, L; 28C, F, L; 29A).

Grain sizes (polar measurements given first, followed by equatorial measurements) vary from (7–)8(–11) × (7–)8(–10) μm to (12–)12.6(–13) × (7–)8.6(–9) μm. In some cases, putatively closely related species have pollen almost identical in size and surface sculpture.

Chromosomes

Solt and Wurdack (1980) reported chromosome numbers of $n = 24$ for *Macairea pachyphylla, M. parvifolia,* and *M. stylosa* and of $n = 22$ for *M. spruceana.* I made repeated attempts to count chromosomes of *M. rufescens, M. neblinae,* and *M. theresiae* grown in a greenhouse in Washington, but, although numerous root tips were preserved, suitable mitotic stages were not obtained.

Pollination and Dispersal

Melastomataceae with a floral morphology similar to that of *Macairea,* for example species

→

FIG. 26. Scanning electron micrographs of pollen grains in *Macairea*. **A–C.** *M. lanata* (*L. Williams 14271*). **A.** Lateral view (×2500). **B.** Polar view (×5000). **C.** Polar view (×2500). **D–F.** *M. stylosa* (*Maguire et al. 35605*). **D.** Lateral view (×2500). **E.** Polar view (×5000). **F.** Polar view (×2500). **G–I.** *M. maroana* (*Maguire et al. 41772*). **G.** Lateral view (×2500). **H.** Lateral view (×5000). **I.** Polar view (×2500). **J–L.** *M. duidae* (*Steyermark et al. 124578*). **J.** Lateral view (×2500). **K.** Lateral view (×5000). **L.** Polar view (×2500).

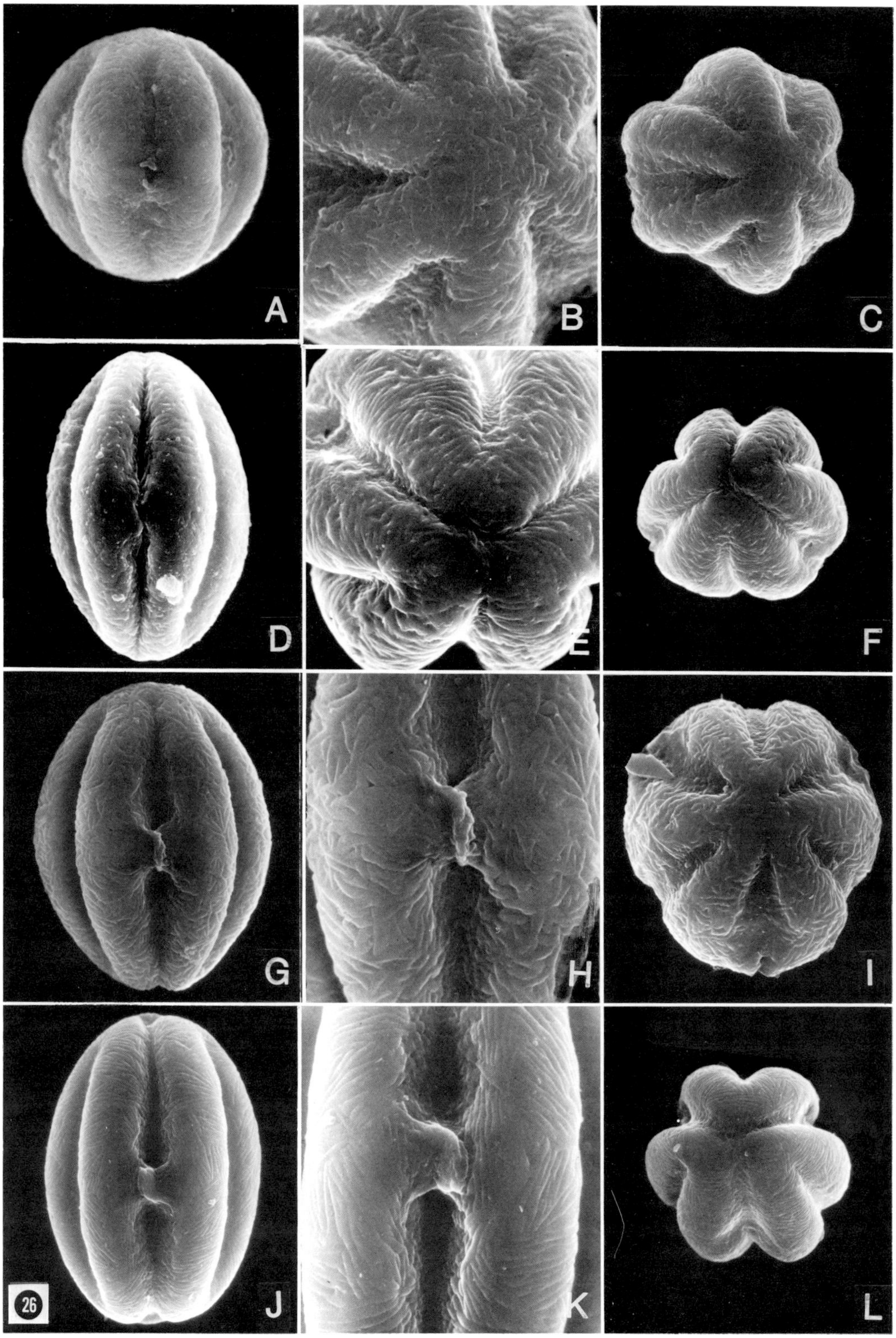
A
B
C
D
E
F
G
H
I
J
K
L
26

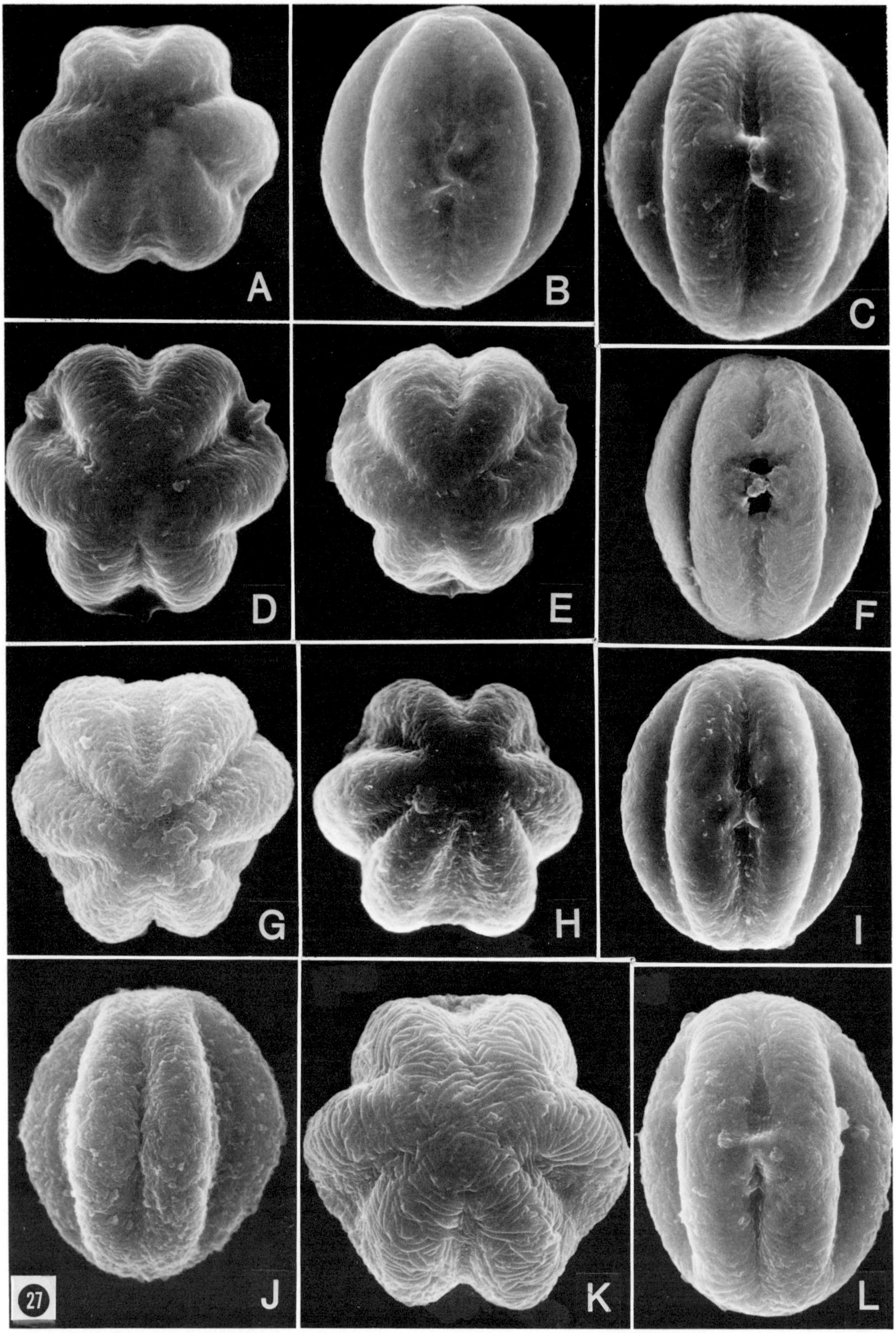
A
B
C
D
E
F
G
H
I
J
K
L
27

of *Tibouchina,* are visited by female bees comprising a wide spectrum of the neotropical bee fauna (Renner, 1984a, 1984b). As a rule, these bees collect pollen (which is the only reward furnished by the flowers) from the poricidal stamens by vibrating them. Bees of the right size contact the stigma in the process and potentially transfer pollen from one flower to another. However, during 40 hours of observation in May, June, and July 1981, 1982, and 1984 of a large population of *Macairea theresiae* near Manaus, I have seen no bee visiting or inspecting the flowers. Unvisited flowers in which the styles were cut off produced viable seeds asexually. Also, most pollen grains in *M. theresiae* did not stain in cotton blue/lactophenol and thus were probably inviable.

In several of the other species, the stamens of herbarium specimens contained few and sometimes no pollen grains. During my SEM survey of the pollen, a high proportion of the grains appeared malformed, and therefore I suspect that facultative apomixis may be widespread in the genus. Agamospermy in melastomes in general, and in those of savanna habitats in particular, is not uncommon (Renner, 1984a and in press).

The small, light seeds are shaken out of the capsules by the wind. Dispersal capacity seems to be good, judged by the fact that some species have colonized isolated mountain tops or savannas many miles from each other.

In the greenhouse, the seeds of *Macairea rufescens, M. neblinae, M. theresiae, M. stylosa, M. spruceana, M. pachyphylla,* and *M. parvifolia* germinated after 16–28 days. (Germination times for the latter four species were kindly provided by John Wurdack.)

Ecology and Distribution

Macairea species require ample light and are adapted to acidic and sometimes waterlogged soils, such as are found in various types of savannas in northern South America. The species with the southernmost distribution, *Macairea radula,* is common in the cerrado vegetation of the Brazilian Planalto where it grows in gallery forests at elevations of 500–1000 m; it also occurs in seasonally inundated grass savannas in western Bolivia. The most widespread species, *M. thyrsiflora,* ranges from the states of Amazonas and Bolívar in Venezuela, through Guyana and Brazilian Amazônia, into Mato Grosso and Bolivia and west into Peruvian and Colombian Amazonia. This highly polymorphic species has a broad ecological amplitude, growing from sea level to 1700 m elevation and occurring in natural grass savannas, in rocky areas on the sandstone mountains of Guayana, and in disturbed habitats such as roadsides.

Seven species are strictly adapted to the climate and soil conditions prevalent on mountain summits of southern Venezuelan highlands (Mayr & Phelps, 1967; Steyermark, 1979): *Macairea cardonae, M. chimantensis, M. duidae, M. linearis, M. neblinae, M. parvifolia,* and *M. rigida* occur only between 1250 and 2500 m elevation. All sandstone mountains on which *Macairea* species have been found and the elevations at which the species grew are given in Figure 30; of the roughly 40–45 sandstone mountains that have been explored botanically as of 1986, 32 (counting the Chimantá and Auyán-tepuí mountain ranges each as a single botanical area) have yielded *Macairea* collections. Two closely related species, *M. parvifolia* and *M. rigida,* apparently replace each other on the eastern and western mountains, respectively. The greatest diversity is found on Mt. Duida, which harbors five species of *Macairea.*

Four species (*Macairea lanata, M. multinervia, M. pachyphylla,* and *M. thyrsiflora*) are widespread in Amazonian lowlands and in the Gran Sabana where they occur at elevations up to 1700 m. Another seven species are found exclusively in lowland savannas (*M. rufescens, M. theresiae,*

←

FIG. 27. Scanning electron micrographs of pollen grains in *Macairea.* **A, B.** *M. sulcata* (*Cuatrecasas 7197*). **A.** Polar view (×2500). **B.** Lateral view (×2500). **C, D.** *M. spruceana* (*Maguire et al. 41832*). **C.** Lateral view (×2500). **D.** Polar view (×2500). **E, F.** *M. pachyphylla* (*Wurdack et al. 2844*). **E.** Polar view (×2500). **F.** Lateral view (×2500). **G, J.** *M. chimantensis* (*Huber 9243*). **G.** Polar view (×2500). **J.** Lateral view (×2500). **H, I.** *M. linearis* (*Steyermark 58142*). **H.** Polar view (×2500). **I.** Lateral view (×2500). **K, L.** *M. multinervia* (*Teixeira et al. 1327*). **K.** Polar view (×2500). **L.** Lateral view (×2500).

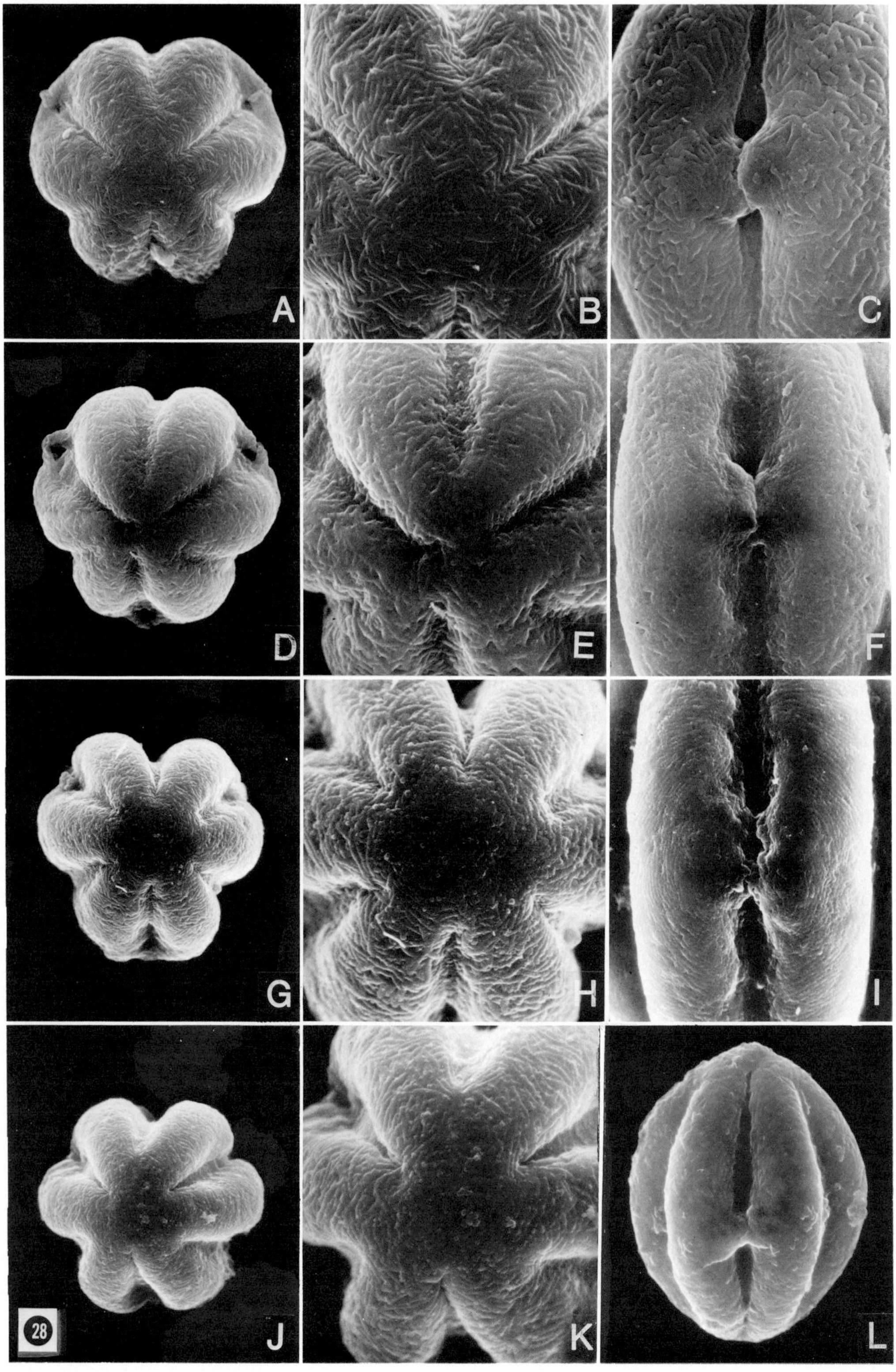
A
B
C
D
E
F
G
H
I
28
J
K
L

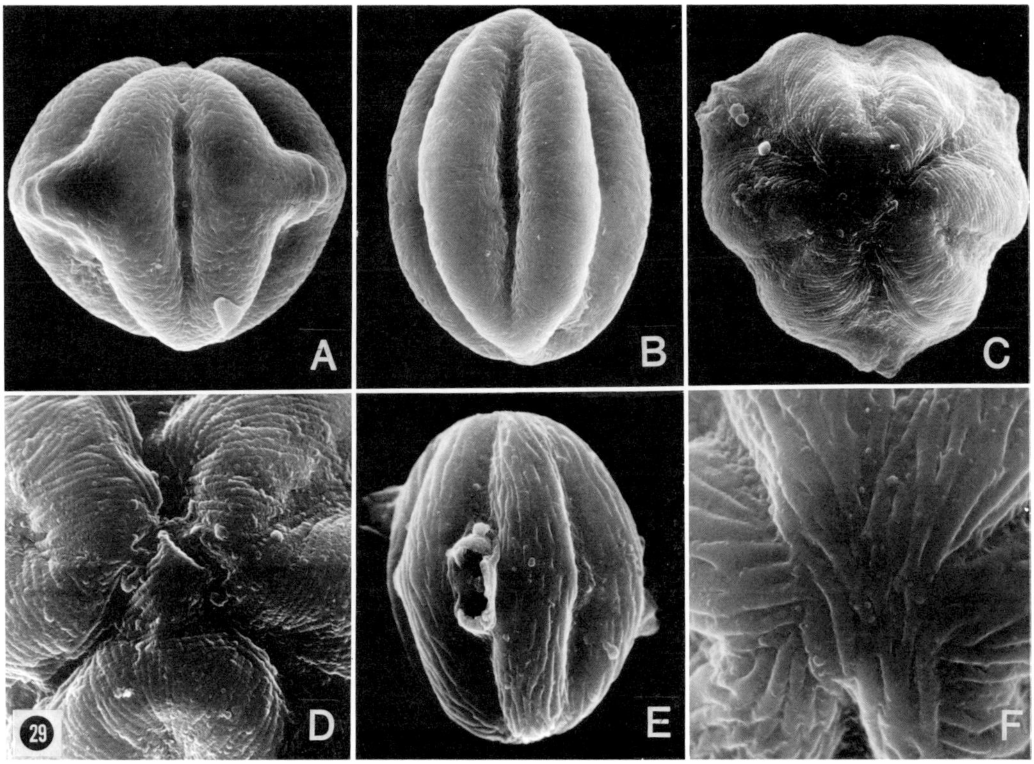

FIG. 29. Scanning electron micrographs of pollen grains in *Macairea*. **A–D.** *M. rufescens* (*Renner 2106*). **A, B.** Lateral views (×2500). **C.** Polar view (×2500). **D.** Polar view (×5000). **E, F.** *M. lasiophylla* (*Maguire & Maguire 40360*). **E.** Lateral view (×2500). **F.** Polar view (×5000).

M. maroana, M. cuieirasii, M. spruceana, M. stylosa, and *M. axilliflora*). The first two form large colonies on dry white sand savannas and do well on burned sites; *M. maroana* grows at the edges between savanna and forest; *M. cuieirasii, M. spruceana, M. stylosa,* and *M. axilliflora* all prefer sandy banks of black-water rivers, where they are often partly or completely submersed for long periods.

From these habitat preferences it is clear that, while a few species of *Macairea* are widespread, the great majority have a limited range and the distribution of all is fragmented. This fact must be borne in mind when reading Figure 31, which shows the superposed distributions of all species. *Macairea* is distributed from the states of Amazonas and Bolívar in Venezuela, east through Guyana and Surinam, south through the Amazon basin to central Brazil and northern Bolivia, and west into Peruvian and Colombian Amazonia; apparently it does not extend into French Guiana. The genus ranges from 8°N to 24°S.

By far the highest concentrations of species occur in the Orinoco-Rio Negro drainage and in the Gran Sabana in Venezuela, particularly in the area of the Chimantá Massif (Fig. 31). In these two centers of *Macairea* diversity 17 of the 22 species are represented. Ten species are lim-

←

FIG. 28. Scanning electron micrograph of pollen grains in *Macairea*. **A–C.** *M. rigida* (*Maguire & Maguire 35068*). **A, B.** Polar views (A ×2500, B ×5000). **C.** Lateral view (×5000). **D–F.** *M. radula* (*Mattos & Heringer 352*). **D, E.** Polar views (D ×2500, E ×5000). **F.** Lateral view (×5000). **G–I.** *M. neblinae* (*Maguire et al. 42095*). **G, H.** Polar views (G ×2500, H ×5000). **I.** Lateral view (×2500). **J–L.** *M. thyrsiflora* (*Tillett et al. 44909*). **J, K.** Polar views (J ×2500, K ×5000). **L.** Lateral view (×2500).

	dui	chi	thy	sul	pac	par	rig	neb	lin	car	las	mul	lan	phi
Tafelberg					+									
Kaieteur			450		450									
Ayangana			750											
Merume Mts.			+											
Roraima			+											
Pacaraima			1150		800						1000			
Aracá			1200		1200	1200								
Ilú-tepui			1600			1200						1300		
Venamo			+											
Uei-tepui			1350								1200			
Ptari-tepui			1500			1500								
Chimantá		1950 2300								1950 2500				
Auyán-tepui		1950 2300			1100	+				+				
Cerro Venado			1300			1250								
Uaipán-tepui		1600 1900	1700		700 1100	1200								
Guaiquinima						750 1800								
Parú					800		800 2000						1600	
Asisa					300		+							
Yutajé					1250		1300							
Moriche			300 1250				1500						800	
Camani			600											
Duida	1200 2100				1250		1600		1700 2000				1500	
Huachamacari					400									
Marahuaca							1500							
Sipapo							+							
Yapacana			1200											
Aracamuni			1550										750	
Cerro Vinilla			400								400			
Unturan			350											
Neblina								1600 2200						
Cerros at Mitú			+	450										
Serranía de la Macarena														+

ited to the Orinoco-Rio Negro drainage and the western cerros (*Macairea duidae, M. spruceana, M. rufescens, M. maroana, M. rigida, M. neblinae, M. linearis, M. lanata, M. axilliflora*, and *M. stylosa*); three species are exclusive to the Gran Sabana and the eastern tepuís (*M. chimantensis, M. parvifolia, M. cardonae*). Three widespread species, *M. thyrsiflora, M. pachyphylla*, and *M. lasiophylla*, occur in both these centers of diversity. The savannas of the middle and lower Amazon basin and the lower Rio Negro harbor only two endemic species of *Macairea, M. theresiae* and *M. cuieirasii.*

On the western outliers of the Roraima sediments, the Serranía de la Macarena and the sandstone terraces along the rivers Caquetá (near Araracuara) and Vaupés (near Mitú), two poorly known endemics occur, *Macairea philipsonii* and *M. sulcata.* The latter has a disjunct distribution: besides near Araracuara and Mitú, it occurs in savannas near Tarapoto in Peru, 600 km away (Fig. 31). Even more impressive is a disjunction of 1200 km in the range of *M. multinervia*, which is known from the Gran Sabana and savannas on the upper Rio Madeira in Brazil. Such disjunctions are likely to be reduced by further collecting.

Inter- and Intrageneric Relationships

The genus *Macairea* is a monophyletic unit characterized by the unique dorsi-basal connective appendage and the presence of glandular hairs on the vegetative and reproductive parts of the plants. It belongs in the Tibouchineae (defined by their cochleate or subcochleate seeds with a circular hilum) where it is closest to the large genus *Tibouchina.* In particular, Cogniaux' *Tibouchina* sections *Diotanthera* and *Simplicicaules* possess characters also found in certain species of *Macairea* such as (1) an abundant pubescence with various hair types, for example, appressed, conical hairs on the upper leaf surface and erect, basally broad, roughened hairs on the lower surface ("Zottenhaare" or "shaggy hairs"; Gleason, 1950b; Pflaum, 1897; Wurdack, 1986); (2) often 4-merous flowers in terminal paniculate cymes with regularly 4-locular ovaries; (3) sericeous or 4-lobed ovary apices; (5) often purple flowers with white centers which change to red on the second day of anthesis; (6) sometimes filaments with short-stalked, globose glands on the ventral side; (7) rostrate anthers; and (8) a shrubby habit. The classical workers Bonpland (1820), De Candolle (1828a, 1828b), Bentham (1840), Baillon (1877), and Cogniaux (1885, 1891a) all regarded *Macairea* as close to *Tibouchina* and in my analysis of the relationships within *Macairea* I have used the sections *Diotanthera* and *Simplicicaules* of *Tibouchina* as outgroups. It is possible that the more than 250 species of *Tibouchina* form a paraphyletic group; the problem of circumscribing *Tibouchina* is confounded by several medium-sized genera, for instance *Brachyotum, Desmoscelis, Pterolepis*, and *Pterogastra*, which appear to be segregates of the larger group.

Within *Macairea* two main groups may be distinguished (Fig. 32): one, subsequently referred to as group A, comprising eight species, with at least two hair types (long eglandular hairs and sessile glands), and another (group B) of 13 species that have only one hair type (either long glandular hairs or, in one species, minute sessile glands not homologous to the sessile glands in the other lineage, or else papillae-like hairs). The glabrous species *M. linearis* seems to belong to the second lineage for reasons given below. Compared to the ancestral condition, the development of different types of hairs and glands on the vegetative *and* reproductive parts of the plants is derived; therefore, characters 1 (only one hair type) and 2 (two types of hairs) in Figure 32 are both derived states defining two sister clades in *Macairea.* The basic evolutionary divergence between the two lineages of *Macairea* seems to have been maintenance of diversity in hair types versus simplification (dedifferentiation) with only one

←

FIG. 30. Sandstone Mountains from which *Macairea* is known. The mountains are listed from east to west, the species in the sequence in which they are treated in the revision (without the lowland taxa). The altitude is given as indicated by the collectors. Dui = *M. duidae*, chi = *M. chimantensis*, thy = *M. thyrsiflora*, sul = *M. sulcata*, pac = *M. pachyphylla*, par = *M. parvifolia*, rig = *M. rigida*, neb = *M. neblinae*, lin = *M. linearis*, car = *M. cardonae*, las = *M. lasiophylla*, mul = *M. multinervia*, lan = *M. lanata*, phi = *M. philipsonii*.

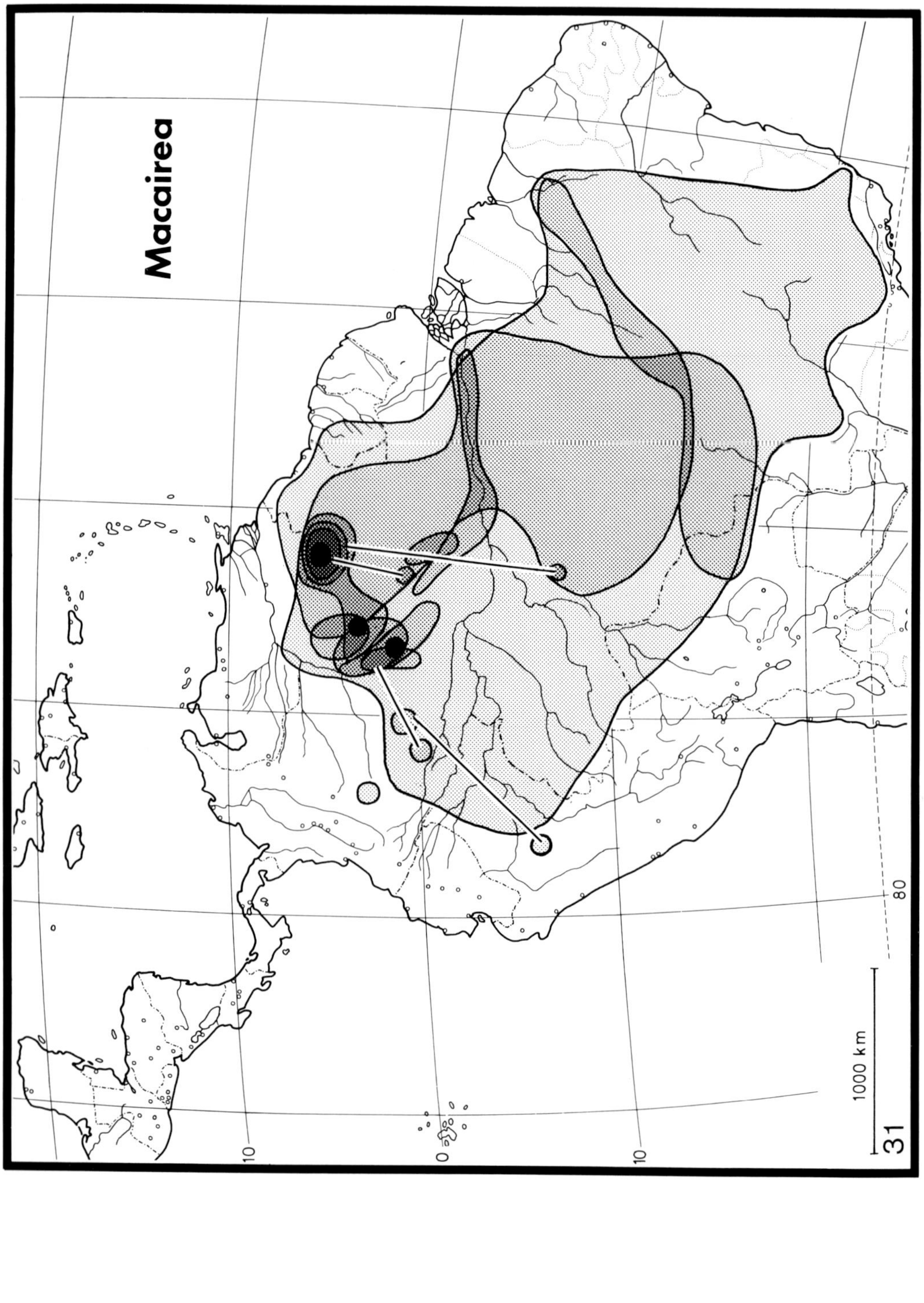

FIG. 31. Distribution of *Macairea*.

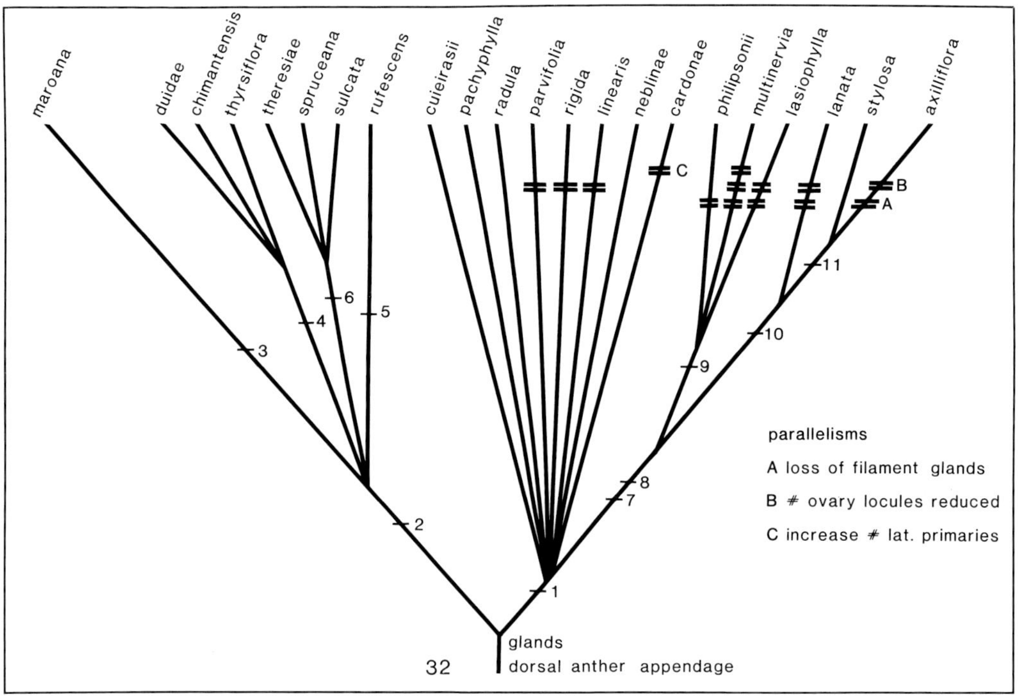

FIG. 32. Cladogram of *Macairea,* see text for discussion. Also shown, the distribution of three selected characters (A, B, C) demonstrating parallel evolution. Numbers refer to the following apomorphies: 1—only one type of hairs on the lower leaf surface and the inflorescence, 2—two types of hairs on the lower leaf surface and the inflorescence, 3—three types of hairs on the lower leaf surface and the inflorescence, 4—erect hairs with greatly roughened bases on the lower leaf surface and inflorescence, 5—leaves pseudo-pinnately veined, 6—ovary apex sericeous with eglandular hairs, 7—congested cymes in the axils of the uppermost foliage leaves, 8—tendency to lose the filament glands, 9—suffrutescent habit, 10—cymes strongly congested and few-flowered, 11—polyhedric seeds.

hair type, i.e., I interpret the 13 species with only one hair type (group B) as showing a more derived condition of the indument than group A.

Though the indument characters do not correlate with other single characters, they do correlate with the following tendencies: in group B, a reduction in the number of ovary locules and a lack of the filament glands are found repeatedly; some species in group B have small cymes or groups of subsessile flowers in the axils of the uppermost vegetative leaves (characters 7 and 10 in Fig. 32; instead of the terminal paniculate cymes found in the outgroups); several species are subshrubs (character 9 in Fig. 32), and some species show an increase in the number of lateral primary nerves. I consider all these characters derived. Group A, on the other hand, has several character states also expressed in the outgroup, such as basally greatly roughened hairs (Fig. 22A–C, character 4 in Fig. 32), sericeous ovary apices with eglandular hairs (character 6), and sometimes 4-lobed ovary apices. Since characters 4 and 6 may really be expressions of plesiomorphic character states (see above, characters of *Tibouchina* sections *Diotanthera* and *Simplicicaules*), they were used reluctantly to unite species in group A (in Fig. 32); it might have been better to leave the two nodes defined by 4 and 6 unresolved.

Most species in group B have long, glandular hairs, but three species placed in this group, *Macairea linearis, M. stylosa,* and *M. axilliflora,* lack such hairs. However, in the morphology of their seeds (Fig. 24A, C) and in the shape and size of their stamens and pistils they are close to other species in group B, particularly to *M. parvifolia* and *M. lanata* (cf. seed of that species, Fig. 24E). In *Macairea stylosa,* the lower leaf surface is covered with minute, few-branched, papillae-like hairs which form a velvety surface (Fig. 23D). Rarely, relatively short, glandular hairs occur in this matrix; except for their reduced length, these glandular hairs are of the same type as the long glandular hairs and therefore, *M. stylosa* is placed in group B. In *M. axilliflora,* the lower leaf surface is covered entirely with minute, sessile glands half the size of the glands found in the species of group A (measurements, see *Indumentum and Glands*) and hardly homologous with them. Though this indument appears very different, it consists of only one type of hair like in all species of group B and, bearing in mind the flower and seed characters (character 11 in Fig. 32), *M. axilliflora* is confidently placed in this group. *Macairea linearis,* finally, has glabrous leaves, which may be rationalized as correlating with its regular occurrence near stream beds where it is often partly submerged. The species is placed among the species with long glandular hairs on the basis of its flowers and seeds.

In *Macairea maroana* (in group A) indument differentiation proceeded further, with different combinations of hair types confined to specific parts of the plant (character 3 in Fig. 32): the leaf undersurfaces, the petioles, the branchlets, and the peduncles bear sessile glands and long eglandular hairs, whereas the pedicels, hypanthia, and calyx lobes have gland-headed long hairs, but neither sessile glands nor long eglandular hairs. Another such autapomorphy (character 5 in Fig. 32) is the pseudo-pinnate venation of *Macairea rufescens.* This type of venation is rare in the Melastomataceae and has arisen repeatedly in unrelated groups, e.g., in species of *Graffenrieda* (Merianeae) and in *Henriettella squamata* Alain (Miconieae); it is considered derived.

The prevalence of losses (e.g., character 8, intraspecific tendency to loose filament glands), reductions (e.g., character 10, congested cymes), and parallelisms (e.g., absence of filament glands, reduced number of ovary locules, increased number of lateral primary veins, indicated as A, B, and C, respectively, in Fig. 32) together with the scarcity of shared derived characters preclude to a great extent postulations of the evolutionary relationships within *Macairea.* Thus, the branching diagram (Fig. 32) is highly unresolved; it was arrived at manually, since it would have been a waste of time letting a computer program switch around all the undistinguishable taxa and produce hundreds of equally parsimonious trees for 22 terminal taxa.

Macairea might be termed a "Guayana genus," i.e., one diversified chiefly in the Guayana phytogeographical province (Maguire, 1970). This phytogeographical unit is defined as the flora on the sandstone deposits and the related sediments (Roraima Formation) that are superposed over the crystalline Guayana Shield. The ancestors of the seven species of *Macairea* endemic on the summits probably established themselves there at various times in the past after their seeds were transported upward by air currents. Sudden, very strong winds are common on the tabletop mountains (J. Wurdack and O. Huber, pers. comm.; pers. obs.). Since the two most widespread lowland species, *M. thyrsiflora* and *M. pachyphylla,* have broad ecological tolerances and are able to grow under the conditions prevalent at mid-elevations (700–1600 m) and since these two species are morphologically close to the summit species, it seems likely that they gave rise in the past to the summit endemics. Genetic drift may have played an important role in the initially small founder populations in recently colonized, isolated savannas; also, if facultative agamospermy (found in at least one species of the genus, cf. POLLINATION AND DISPERSAL) is widespread in *Macairea,* this may have allowed rapid build-up of homogeneous populations (clones) from single propagules. Certainly, the high incidence of endemism in *Macairea* is correlated with the occurrence of most of the species in small, isolated savannas in an area of relatively great topographical diversity.

TAXONOMIC TREATMENT

Generic Description

Macairea De Candolle, Prodr. **3:** 109. 1828a. Type: *Macairea radula* (Bonpl.) De Candolle. *Rhexia radula* Bonpland *in* Humboldt & Bonpland, Monogr. Melast. **2:** 107, pl. 41. 1820.

Siphantheropsis Brade, Inst. Nac. Pesq. Amazonia Publ. Bot. **8:** 4. 1958. Type: *Siphantheropsis williamii* Brade.

Shrubs, 1–2 m tall, *subshrubs,* or rarely *treelets* (5–12 m tall); branchlets angled; bark longitudinally fissured, exfoliating in thin strips; indument on the young branchlets densely hirsute, sericeous, or lepidote (in *M. linearis*) with glandular or eglandular hairs, sometimes with globose sessile glands, rarely glabrous. *Leaves* opposite, elliptic, broadly ovate, obovate, cordate, or linear, petiolate, 3–9-nerved, the lateral pairs of primary veins arising at the blade base or up to 4 mm above the base, the outermost pair close to the margin and often inconspicuous (leaves may appear almost pinnately veined), subcoriaceous or thin and brittle, upper surface drying green, with stout, conical, glandular or eglandular hairs, or verrucose, or glabrous, lower surface rust-colored or pale green, hirsute or sericeous with glandular or eglandular hairs, sometimes with globose sessile glands, or with very short few-branched papillae-like hairs, or glabrous. *Inflorescence* a terminal, many-flowered elongate or congested paniculate cyme, or groups of 3–5 subsessile flowers in the axils of the uppermost leaves; peduncles hirsute or sericeous with glandular or eglandular hairs, sometimes with globose sessile glands, or lepidote; pedicels with bracts and bracteoles, indument usually of the same type as on the peduncle; bracts subpersistent, margin often hyaline and serrulate. *Flowers* hermaphroditic, 4-merous; hypanthium campanulate, terete or inconspicuously 8-ribbed, hirsute or sericeous with indument of the same type as the pedicel and bracts, or glabrous; calyx lobes narrowly to broadly triangular, spreading, persistent on the limb after anthesis; petals thin, obovate, apically sometimes emarginate, basally with a minute claw, basally white, turning red with age, apically orchid rose, pink, or magenta, or petals white, or purple, margins (at least in bud) with globose short-stalked glands; stamens eight, dimorphic or subisomorphic, of roughly the same length as the petals; filaments thin, terete, ventrally with globose short-stalked glands or glabrous; anthers linear, single-pored, rostrate, rarely truncate, connective prolonged and expanded dorsi-basally into a cordate or collar-like appendage often surrounding the insertion of the filament; ovary free, perigynous, 2–4-locular, apically with glandular or rarely eglandular hairs; style thin, terete or subclavate (*M. lasiophylla*), sigmoid, in the mid-section with short glandular hairs, longer than the stamens and exserted from the corolla; stigma punctiform. *Fruit* a loculicidal, many-seeded, dark brown capsule. *Seeds* 0.5–1.4 mm long, cochleate or polyhedric, testa tuberculate or reticulate-foveate.

A neotropical genus of 22 species, ranging from the Guyana Highlands through Venezuelan and Brazilian Amazonia to central Brazil and northern Bolivia and west into Colombia and Peru; savannas in lowlands and on the Guayana sandstone mountains to 2500 m.

Key to the Species of *Macairea*

1. Flowers subsessile in the axils of the uppermost leaves; petals 3–4 mm long. 22. *M. axilliflora.*
1. Flowers pedicellate, borne in a pedunculate inflorescence; petals longer than 4 mm.
  2. Lower leaf surface glabrous or with minute, papillae-like hairs, without sessile glands.
    3. Leaf blades narrowly elliptic, 3–12 × 1–2.5 cm. 21. *M. stylosa.*
    3. Leaf blades linear, 1.2–1.8 × ca. 0.2 cm. 15. *M. linearis.*
  2. Lower leaf surface with either long, glandular hairs or globose, sessile glands.
    4. Lower leaf surface with long, eglandular hairs, with sessile glands.
      5. Inflorescence with long, glandular hairs. 8. *M. maroana.*
      5. Inflorescence without glandular hairs.
        6. Ovary apically sericeous with long, eglandular hairs.
          7. Leaf blades obovate, apically rounded, with two pairs of lateral primary veins, the lateral secondary veins and the veinlets raised on the lower surface and conspicuous. 4. *M. theresiae.*
          7. Leaf blades obovate or elliptic, apically obtuse or apiculate, with one marginal pair of lateral primary veins, the lateral secondary veins and the veinlets not raised on the lower leaf surface.
            8. Petioles 0.6–1.2 cm long; blades 2.5–4 cm wide, apically cuspidate. 6. *M. sulcata.*
            8. Petioles to 0.5 cm long; blades 1.5–2.5 cm wide, apically obtuse. 5. *M. spruceana.*
        6. Ovary apically with few, short, glandular hairs.
          9. Leaf blades pseudo-pinnately veined, the single pair of lateral primary veins inconspicuous; upper surface glabrous and smooth, lower surface sparsely hirsute with long, prostrate, subulate hairs which are apically frayed. 7. *M. rufescens.*

9. Leaf blades with 1–2 pairs of conspicuous lateral primary veins; upper surface verrucose or scabrous, with stout, conical hairs, lower surface with erect, basally broad and roughened hairs.
10. Leaf blades narrowly elliptic, 3.5–5.5 × 1.8–2.5 cm; calyx lobes 1–2 mm long; filaments 2.8–3.2 or 1.8–2 mm long, thecae 1.5–2 mm long. 2. *M. chimantensis.*
10. Leaf blades broadly elliptic or elliptic or obovate, 3.5–13 × (2–)2.5–4.5 cm; calyx lobes to 5 mm long; filaments 4–9 or 3–8 mm long, thecae 3–5 mm long.
11. Leaf blades 3.5–6 × 2.5–4 cm; calyx lobes 1–2 mm long; filaments 3.8–4.2 or 2.8–3.3 mm long. 1. *M. duidae.*
11. Leaf blades 4.5–13 × (2–)3–4.5 cm; calyx lobes 2–5 mm long; filaments 8–9 or 5–8 mm long. 3. *M. thyrsiflora.*
4. Lower leaf surface with long, glandular hairs, lacking sessile glands.
13. Cymes in the axils of the uppermost foliage leaves. 17. *M. lanata.*
13. Cymes terminal.
14. Leaf blades with 3–4 pairs of lateral primary veins; petals 14–15 × 10–15 mm.
15. Petioles 0.5–1.5 cm long; inflorescence congested; filaments 5–6 or 4–5 mm long, with short-stalked glands; ovary 4-locular. 16. *M. cardonae.*
15. Petioles 0.3–0.8 cm long; inflorescence elongate; filaments 8–10 or 7–8 mm long, glabrous; ovary 3(–4)-locular. 19. *M. multinervia.*
14. Leaf blades with 1–2 pairs of lateral primary veins; petals to 9 mm long, to 6(–8) mm broad.
16. Petioles 0.3–0.5 cm long; filaments glabrous.
17. Pedicels 2–3 mm long; calyx lobes 1.5–2.3 mm long; ovary 3-locular; petals basally white, turning red with age, apically purple. 20. *M. lasiophylla.*
17. Pedicels 4–7 mm long, calyx lobes 3–4 mm long; ovary 4-locular; petals purple. 18. *M. philipsonii.*
16. Petioles longer than 0.5 cm; filaments with short-stalked glands.
18. Ovary 3(rarely 2)-locular.
19. Leaf blades elliptic, 2–4 × 0.8–1.8 cm. 12. *M parvifolia.*
19. Leaf blades obovate, 4–7 × 2–4 cm. 13. *M. rigida.*
18. Ovary 4-locular.
20. Leaf blades narrowly elliptic, (3.5–)6–8 × 2–3.5 cm. 14. *M. neblinae.*
20. Leaf blades elliptic or obovate, (4.5–)7–15 × (1.8–)3(–9) cm.
21. Lower leaf surface densely sericeous with long, slightly curled and short, erect, glandular hairs, lateral secondary veins outlined with dense pubescence of long hairs; thecae 4–5 mm long; filaments 10–11 or 8–9 mm long. 11. *M. radula.*
21. Lower leaf surface hirsute with only short, glandular hairs; thecae 2–4 mm long; filaments 5–8.5 or 3–6.5 mm long.
22. Leaf blades with 1 pair of lateral primary veins, the lateral secondary veins 3–6 mm apart; hypanthium 2.8–3.2 × 3.1–3.7 mm; thecae ca. 4 mm long. 9. *M. cuieirasii.*
22. Leaf blades with 2 pairs of lateral primary veins, the lateral secondary veins 2–4 mm apart; hypanthium 1.5–2 × ca. 2 mm; thecae 2(–3) mm long. 10. *M. pachyphylla.*

1. Macairea duidae Gleason, Bull. Torrey Bot. Club **58:** 416. 1931b. Type: Venezuela. Terr. Fed. Amazonas: Mt. Duida, Savanna Hills, 1500 m, Aug 1928–Apr 1929 (fr), *Tate 1024* (holotype: NY!, NY photo neg. s.n.; isotype: US!). Figs. 24G, H; 26J–L.

Shrub, 1.5–2 m tall; branchlets, petioles, and peduncles strigose with slender, basally broadened hairs and globose sessile glands. *Leaves* broadly elliptic or obovate, 3.5–6 × 2.5–4 cm, obtuse or minutely apiculate, rounded at base, 3(–5)-nerved, the lateral primaries arising at the blade base, the outermost pair close to the margin, upper surface scabrous, with short, stout, conical hairs, lower surface on the mid-vein and lateral primary veins strigose with basally broadened hairs, the lateral secondary veins and veinlets hirsute with subulate hairs, areoles with globose sessile glands; petioles 0.3–0.7 cm long. *Inflorescence* a many-flowered terminal paniculate cyme; pedicels, bracts, and hypanthium strigose with subulate eglandular hairs, with sessile glands. *Flowers* with pedicels 1–2 mm long; hypanthium 1.8–2.2 × 1.7–2.1 mm; calyx lobes narrowly triangular, glabrous, to 2 mm long; petals dull orchid rose, 5–6 × 3–4 mm; filaments

3.8–4.2 or 2.8–3.3 mm long, ventrally with short-stalked glands; thecae ca. 3 mm long, the connective prolonged and expanded dorsi-basally; ovary apically with short glandular hairs, 4-locular. *Fruit* capsular. *Seeds* 0.8–0.9 mm long, subcochleate.

Distribution and ecology (Fig. 33B): Known only from Mt. Duida at elevations between 1230 and 2100 m; growing on ridge tops or in *Heliamphora*-dominated swamp savannas.

Representative specimens examined. VENEZUELA. AMAZONAS: Plateau of Mt. Duida above Caño Culebra, 3°36′N, 65°42′W, 1250 m, *Liesner 18162* (MO, US); 2100 m, *Farinas et al. 551* (NY, VEN); 1500–1600 m, *Maguire et al. 29644* (NY, US); 1230 m, *Steyermark et al. 126398* (NY, US, VEN); 1500 m, *Tate 1024* (NY, US).

2. **Macairea chimantensis** Wurdack, Mem. New York Bot. Gard. **10:** 138. 1964. Type: Venezuela. Bolívar: Frequent in lower part of lower northwest cumbre, Churí-tepuí (Muru-tepuí), Chimantá Massif, 2100–2200 m, 24 Jan 1953 (fl), *Wurdack 34177* (holotype: US!; isotypes: G!, NY!, P!). Figs. 22D; 27G, J.

Shrub or *treelet,* 0.3–7 m tall; branchlets, petioles, and peduncles strigose with slender, basally broadened hairs and globose sessile glands. *Leaves* narrowly elliptic, 3.5–5.5 × 1.8–2.5 cm, obtuse, acute at base, 3-nerved, the lateral pair of primaries arising at the blade base and close to the margin, upper surface verrucose, lower surface on the mid-vein, the lateral primary veins and lateral secondary veins strigose with short, erect, basally broad and roughened hairs, the veinlets hirsute with subulate hairs, mid-vein and areoles with sessile glands; petioles 0.8–1.5 cm long. *Inflorescence* a many-flowered terminal paniculate cyme; pedicels, bracts, hypanthium, and calyx lobes subsericeous with eglandular hairs and with sessile glands. *Flowers* with pedicels 1–2 mm long; hypanthium 1.8–2.2 × 1.7–2.1 mm; calyx lobes triangular, 1–2 mm long; petals basally white, apically pink, 8–10 × 7–7.5 mm; filaments 2.8–3.2 or 1.8–2 mm long, ventrally with short-stalked glands; thecae 1.5 mm long, connective minutely prolonged and expanded dorsi-basally; ovary apically with short glandular hairs, 4-locular. *Fruit* capsular. *Seeds* 0.9–1 mm long, elongate.

Distribution and ecology (Fig. 33B): *Macairea chimantensis* is known from exposed areas on the mountains of the northeastern Gran Sabana (i.e., the Chimantá Massif, Uaipán-tepuí, Auyán-tepuí and the Aparamán range just east of Auyán-tepuí); at elevations from 1600 to 2300 m.

Representative specimens examined. VENEZUELA. BOLIVAR: Summit of Cerro Apacará, Río Caroní, *Cardona 1555* (US); summit of Auyán-tepuí, 2100 m, *Cardona 2652* (NY, US, VEN); Chimantá Massif, Churí-tepuí, 5°15′N, 61°58′W, 2250 m, *Huber 9243* (US); Chimantá, 5°18′N, 62°09′W, ca. 2000 m, *Steyermark et al. 128128* (MO, US, VEN); Auyán-tepuí, 2200 m, *Tate 1148* (NY, US, VEN); Churí-tepuí (Muru-tepuí), 2100–2200 m, *Wurdack 34177* (B, G, NY, US).

As noted in the protologue, the species seems closest to *Macairea thyrsiflora* (syn. *M. aspera*) from which it differs in the smaller and narrower leaves and the smaller flowers and fruits.

3. **Macairea thyrsiflora** De Candolle, Prodr. **3:** 109, 484. 1828a. Type: Brazil. Amazonas: Upper Rio Negro (fl), *Ferreira s.n.* (holotype: P!; isotypes: K!, LISU!). Figs. 22A–C; 23A; 28J–L.

Macairea albiflora Cogniaux *in* Martius, Fl. bras. **14(3):** 246. 1885. Type: Colombia (Vichada or Guainía) or Venezuela (Amazonas). Nr. Maypures or on the Atabapo, Jun 1854 (fl), *Spruce 3719* (holotype: K!, H. A. Gleason photo s.n.; isotypes: BM!, K!, P!, F photo neg. 36139!, W!, F photo neg. 32323!).

Macairea aspera N. E. Brown, Trans. Linn. Soc. London **6:** 27, pl. 2. 1901. Type: Guyana. Upper slopes of Mt. Roraima, 1894 or 98 (fl), *McConnell & Quelch 31* (holotype: K!).

Macairea scabra Cogniaux, Bot. Jahrb. Syst. **42:** 133. 1908 [1909]. Type: Peru. Loreto: Rioja, W of Moyobamba, 800–900 m, Sep 1904 (fl), *Weberbauer 4708* (holotype: BR!; isotypes; G!, isotype at B destroyed, but represented by a photo, F neg. 16704!).

Macairea rosea Cogniaux, Comm. Linh. Telegr. Estr. Matto Grosso, Publ. 10, Annexo 5, Bot. **3:** 4. 1912. Type: Brazil. Mato Grosso: Juruena, Apr 1909 (fl), *Hoehne 1800* (lectotype, here designated: BR!, photo US s.n.!).

Macairea arirambae Huber, Bull. Soc. Bot. Genève **6:** 193, fig. 6. 1914. Type: Brazil. Pará: Campos de Ariramba, Trombetas, 23 Dec 1906 (fl), *Ducke 8087* (lectotype, here designated: MG!, F photo neg. 2602!; isolectotypes: G!, MG!).

Macairea villosa Hoehne, Mem. Inst. Butantan, Sec. Bot. **1:** 59, pl. 8, fig. 1. 1922. Type: Brazil. Mato Grosso: Morro Podre, Chapada Central, Mar/Apr 1911 (fl), *Hoehne 2476* (holotype: R!).

Macairea schultesii Wurdack, Bot. Mus. Leafl. **18:** 164. 1958a. Type: Colombia. Comisaria del Vaupés: Río Kuduyari, Cerro Yapoboda, savannas on quartzite base, 450 m, 5/6 Oct 1951 (fl), *Schultes & Cabrera 14358* (holotype: NY!).

Macairea maguirei Wurdack, Phytologia **20(6):** 36.

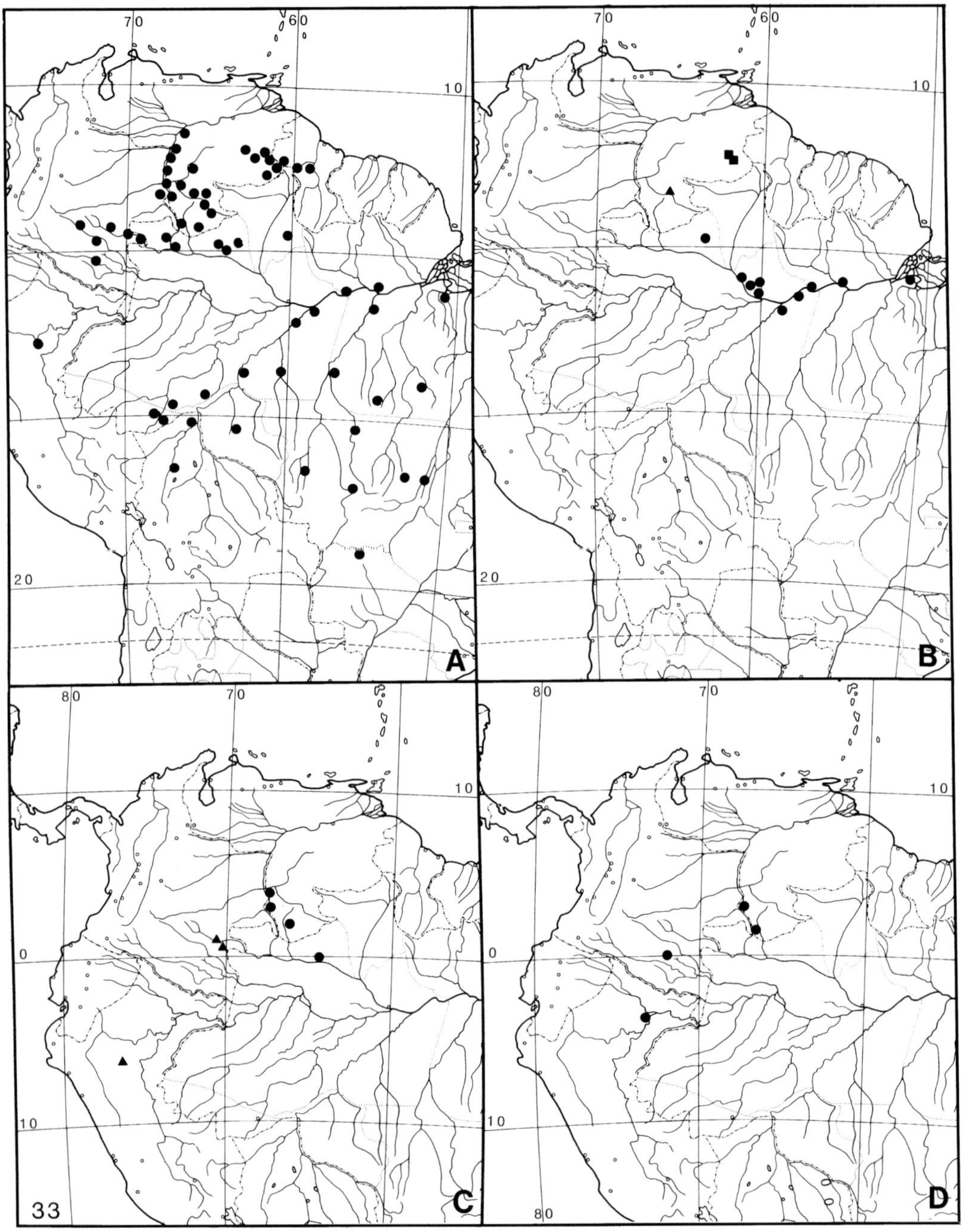

FIG. 33. Distribution of *Macairea*. **A.** *M. thyrsiflora*. **B.** *M. theresiae* (●), *M. duidae* (▲), *M. chimantensis* (■). **C.** *M. spruceana* (●), *M. sulcata* (▲). **D.** *M. rufescens*.

1970. Type: Venezuela. Terr. Fed. Amazonas: Edges of sabana El Venado, left bank of Caño Pimichín, 2 km above Pimichín, 120–140 m, 10 Oct 1957 (fl), *Maguire, Wurdack & Keith 41800* (holotype: US!; isotypes: B!, G!, NY!, P!, VEN!).

Shrub or *treelet*, 0.5–2(–8) m tall; branchlets, petioles, and peduncles sericeous or strigulose with slightly curled, rust-colored, basally roughened hairs and with globose sessile glands. *Leaves* elliptic, 4.5–13 × (2–)3–4.5 cm, obtuse, acute or acuminate, basally obtuse, attenuate or subacute, 3(–5)-nerved, the lateral primaries arising at the blade base, basally close to or to 6 mm from the margin, apically close to the margin, upper surface verrucose or scabrous, with stout conical hairs with downward curved tips, lower surface on the mid-vein and lateral primary veins more or less densely strigose with basally broad and roughened, apically slender, slightly curled hairs, the lateral secondary veins and veinlets with shorter hairs, areoles with short, few-branched papillae-like hairs and with sessile glands; petioles (0.2–)0.5–1.6 cm long. *Inflorescence* a many-flowered terminal paniculate cymes. *Flowers* with pedicels 1–4 mm long; pedicels, bracts, hypanthium, and calyx lobes sericeous with rust-colored, slightly curled hairs, with sessile glands; hypanthium 3–4 × 2–3 mm; calyx lobes broadly to narrowly triangular, 2–3–5 mm long; petals basally white, turning red with age, apically pink, or petals white, 7–9 × (4–)6.5–8 mm; filaments 8–9 or 5–8 mm long, ventrally with short-stalked glands; thecae 3.5–5 mm long, the connective distinctly prolonged and expanded dorsi-basally; ovary apically with short glandular hairs, sometimes with 4 lobes surrounding the base of the style, 4-locular. *Fruits* capsular. *Seeds* 0.7–0.8 mm long, subcochleate.

Distribution and ecology (Fig. 33A): *Macairea thyrsiflora* occurs from the Gran Sabana in Venezuela, east into Guyana, south through Brazilian Amazonia into Mato Grosso and Bolivia, and west into Venezuelan, Colombian, and Peruvian Amazonia. It grows in humid locations, sometimes with the roots in water and is common in disturbed areas; lowlands to 1700 m.

Representative specimens examined. COLOMBIA. AMAZONAS: Río Caquetá, Araracuara, *Idrobo et al. 11234* (COL). CAQUETA: Macaya-Ajaju river confl., Mt. Chiribiquete, 570–700 m, *Schultes 5489* (US). GUAINIA: Puerto Irúrida, sabanas del Aeropuerto, Rio Irúrida, 400 m, *García Barriga 20869* (F, US). VAUPES: Mitú, lower Río Paraná-Pichuna, *Zarucchi 2019* (COL, MO, US). VICHADA: Río Mataveni (confl. del Río Orinoco) ca. 2 km above mouth (nr. Las Palmas), 1–140 m, *Bunting et al. 3602* (US).

VENEZUELA. AMAZONAS: Serranía Parú, Cano Asísa, Río Ventuari, occ. in cumbre along W rim just N of camp caño head, 2000 m, *Cowan & Wurdack 31067* (NY, VEN); La Esmeralda, *Croizat 113* (F, US, VEN); 18 km S of Samariapo toward Río Sipapo, rd. side, 125 m, *Gentry & Berry 14619* (MO, NY, US); Dept. Atures, cuenca del Río Manapiare, pie del Cerro Morrocoy, a 12 km al W de San Juan de Manapiare, 05°19′N, 60°06′W, *Huber 1173* (NY, US, VEN); Dept. Atabapo, pie oriental del Cerro Cucurito, ribera sur del medio Caño Yagua, 3°36′N, 66°34′W, 120 m, *Huber 3880* (US, VEN); between Esmeralda and base of Duida, 175 m, *Maguire & Wurdack 34680* (F, NY, US, VEN); Cerro Yapacana, 1200 m, *Maguire et al. 30668* (NY, US, VEN). BOLIVAR: Slopes of Ilú-tepuí, 1600 m, *Maguire 33286* (NY); Ptarí-tepuí, 1500 m, *Maguire & Wurdack 33824* (NY); head of Salto Pitón, Cerro Pitón, Cordillera Epicara, 400 m, *Maguire et al. 53663* (B, F, NY, US, VEN); Uaipán-tepuí, 1700 m, *Phelps 399* (NY); Sta. Teresita de Kavanayén, NW to Río Karuai, 1220 m, *Steyermark 59356* (F, NY).

GUYANA. Nr. Kaieteur Falls, 460 m, *Cowan & Soderstrom 1721* (F, NY, US); Kaieteur Falls, Mt. Ayanganna, 400 m, *Maguire et al. 40696* (F, NY, P, US); Upper Mazaruni Distr., ca. 2 km S of Waramadan, *Renz 14327* (U).

PERU. LORETO: Lagunas, Pasto Grande (Huallaga river), *McDaniel 16457* (F, MO, NY, US). SAN MARTIN: Río Negro, *Woytkowski 6189* (G, MO, R, US).

BRAZIL. AMAZONAS: Rio Curicuriarí, affl. do Rio Negro, *Ducke 352* (MO, NY, R, S, US); Humaitá, Rio Madeira, *Ducke s.n. (RB 35616)* (G, S, US); Manaus–Caracaraí rd., km 585, *Lima 461* (INPA); Rio Içana, Tunuí, *Lützelburg 22465* (M); along Rio Marié at Manauná, 00°40′S, 66°45′W, *Poole 1875* (MO, NY, US); Fortaleza savanna, Rio Puciarí, tributary of Rio Ituxi, 20 km above mouth, *Prance et al. 13775* (COL, F, M, MO, NY, P, S, US); 3 km S da parte central de Serra Aracá, 8 km E do Rio Jauarí, 0°49′N, 63°19′W, 60 m, *Rodrigues 10473* (INPA, NY, US); Borba on Rio Madeira, 3 mi E of town, 4°23′S, 59°35′W, *Todzia et al. 2220* (MG, US); Rio Padauirí, *Traill 243* (K, P). MATO GROSSO: Chapado dos Guimarães, *Hatschbach et al. 36134* (NY, US); Juruena, *Hoehne 1801* (S); ca. 270 km N of Xavantina–São Felix rd., Base Camp of exp., 12°54′S, 51°52′W, *Ratter 2000* (MO, P, RB, US). PARÁ: Serra do Cachimbo, *Alvarenga s.n. (RJ 90559)* (M, NY, US); Fazenda Santa Olimpia, Mun. Faro, *Black & Ledoux 50-10569* (US); campos de Ariramba (Trombetas), *Ducke 14943* (US); Rio Tapajoz, Bella Vista, *Ducke 17053* (BM, F frag., G, P, RB, US); Rio Cururú, campo da Tabua, *Egler 852* (US); Mun. Oriximiná, campos de Ariramba, Rio Jaramacarú, 70 m, *Martinelli 6882* (NY, RB, US); 10 km E of Portel, *Prance et al. 1288* (NY, US); Serra do Cachimbo, aeroporto, *Rodrigues 9414* (INPA). RONDÔNIA: Porto Velho–Abunã Hwy., 120 km S of Porto Velho, *Calderon et al. 2821* (INPA, US); Alto Rio Branco, way from maloca Buritará to São Luiz, *Scolnik & Luti 755* (NY). RORAIMA: Manaus–Caracaraí rd., km 350, *Rodrigues 10120* (INPA).

BOLIVIA. BENI: Prov. Vaca Diez, *Krapovickas & Schinini 35065* (MO, US). LA PAZ: Prov. Iturralde, Luisita, 13°5′S, 67°15′W, 180 m, *Beck 10072* (LPB, US).

Macairea thyrsiflora is superficially similar to *M. theresiae,* from which it differs in the rust color of the lower leaf surface (vs. a greenish lower surface in *M. theresiae*), the finer reticulation of the veinlets, and the glandular ovary indument (see also under *M. theresiae*). *Macai-*

rea thyrsiflora and *M. theresiae* are sympatric near Oriximiná, Belém, and Santarém; they maintain their distinctness at these localities.

Macairea thyrsiflora is exceedingly polymorphic in the length of the pubescence and the calyx lobes. Specimens from the Río Negro-Pacimoni area have quite smooth leaves with a verrucose upper leaf surface and a velvety lower surface covered with very short, papillose hairs; plants from Guyana have rougher (scabrous) leaves with short, conical hairs with downward curved tips on the upper surface and a more or less strigose lower surface. This east to west cline is found only in the northern part of the species' range. Collections from the area south of the Amazon are homogeneously intermediate in their pilosity. Those from savannas just north of the lower Amazon (Oriximiná and the Campos de Ariramba), in turn, are intermediate between Guyana specimens and the specimens of southern provenance. The quality of the indument is the same in all collections. The length of the calyx lobes ranges from two to five mm and seems to vary randomly.

In some savannas, the rough-leafed and the smooth-leafed morphs occur together; for example, at the base and on the plateau of Serra Aracá, around Esmeralda, near Mitú on the Río Vaupés, and in a savanna at km 350 of the Manaus–Caracaraí road. A savanna colonized by seedlings from several different areas would contain the different plant types growing next to each other for prolonged periods of time if gene exchange between the individuals were low. I suspect this is the case; the pollen grains of *M. thyrsiflora* are often malformed and there are very few grains per anther. In *M. theresiae,* agamospermy has already been demonstrated experimentally (see POLLINATION AND DISPERSAL), and it is possible that the capacity to form seeds asexually is also present in *M. thyrsiflora.* Such offspring would have the advantage of having genotypes already tested for suitability in a particular savanna environment. Asexual reproduction results in clones, with major discontinuities between them and little heterogeneity within them. This would fit well the situation observed in *M. thyrsiflora.*

N. E. Brown (1901) provides an illustration of the species.

4. Macairea theresiae Cogniaux, Bot. Centralbl. **66:** 369. 1896. Type: Brazil. Amazonas: Rio Negro, Manaus, "capoeira," Jul 1888 (fl), *Therese von Bayern s.n.* (holotype: BR frag.!; isotype: M!, US photo neg. s.n.!). Fig. 23C.

Macairea glabrescens Pilger, Verh. Bot. Vereins Prov. Brandenburg **47:** 165. 1905. Type: Brazil. Amazonas: Rio Negro, Manaus, Ponta Negra white sand savanna, May 1902 (fl), *Ule 6153* (holotype at B destroyed, but represented by a photo, F neg. 16699!; isotypes: G-DEL!, HBG!, K!).

Shrub or *treelet,* 2–3 m tall; branchlets, petioles, and peduncles loosely sericeous with apically slightly curled, basally roughened hairs and with globose sessile glands. *Leaves* obovate, 5–14.5 × 1.7–6 cm, rounded or retuse, acute at base, 5-nerved, the lateral primaries arising at the blade base, the outermost pair close to the margin, upper surface verrucose or scabrous, lower surface on the mid-vein and lateral primary veins sericeous with basally roughened, apically slightly curled hairs, with sessile glands, the lateral secondary veins 2–4 mm apart, the veinlets loosely reticulate, sparsely hirsute with thin, often downward curved hairs, areoles with sessile glands; petioles ca. 1.5 cm long. *Inflorescence* a many-flowered terminal paniculate cyme. *Flowers* with pedicels ca. 4 mm long; pedicels, bracts, hypanthium, and calyx lobes sericeous with slightly curled hairs, with sessile glands; hypanthium 2.8–3.2 × 1.8–2.2 mm; calyx lobes broadly triangular, 1.5–2 mm long; petals basally white, turning red with age, apically purple, 7–9 × 6–7 mm; filaments 5.8–6.2 or 3.6–4.2 mm long, ventrally with short-stalked glands; thecae 2–4 mm long, the connective prolonged and expanded dorsi-basally; ovary apically sericeous with eglandular hairs, 4-locular. *Fruit* capsular. *Seeds* 0.8 mm long, subcochleate with dorsal ridges.

Distribution and ecology (Fig. 33B): *Macairea theresiae* is characteristic of the white sand savannas of the lower Rio Negro and the middle and lower Amazon (near Manaus, Oriximiná, Santarém, Belém).

Representative specimens examined: BRAZIL. AMAZONAS: Parintíns, Campo Grande, *Ducke s.n.* (RB, VEN); Manaus–Caracaraí, BR 174, km 130, *Loureiro et al. s.n.* (INPA); 2 km from Rio Cuieiras and 2 km below mouth of Rio Brancinho, *Prance et al. 17931*

(F, MO, NY, S, US); Manaus–Caracaraí rd., km 130, Igarapé Lages, *Prance et al. 21080* (M, MO, NY, P, S, US); Manaus, Ponta Negra, *Renner 284* (HBG, INPA, US); Rio Abacaxis, 4°12′S, 58°40′W, *Todzia et al. 2344* (MG, US). PARÁ: Campina do Jacopí, nr. Gurupá, *Ducke 15956* (BM, G, RB); Gurupá, campina da Grenalda, *Pires & Silva 4722* (NY, UB).

The species is similar in general aspect to *Macairea thyrsiflora*; it differs in its leaves, which are apically rounded (vs. obtuse, acute or acuminate in *M. thyrsiflora*), have five lateral primary veins (vs. 3(–5) in *M. thyrsiflora*), more loosely reticulate veinlets, and lack papillae-like hairs in the areoles. Flowering specimens of the two species may be distinguished by the long eglandular hairs on the ovary of *M. theresiae* vs. the short glandular hairs on the ovary apex of *M. thyrsiflora*. Near Manaus (Amazonas, Brazil), a large population of *Macairea theresiae* was found to produce seeds apomictically (see POLLINATION AND DISPERSAL).

5. Macairea spruceana Berg ex Triana, Trans. Linn. Soc. London **28:** 38. 1871. Type: Brazil. Amazonas: Rio Negro, between São Gabriel and Barcellos, Dec 1851 (fl), *Spruce 2013* (holotype: K!; isotypes: BM!, BR frag.!, G!, F photo neg. 26063!, G-DC!, GOET!, LE!, M!, NY!, OXF!, P!, RB!, W!; B destroyed, but represented by a photo, H. A. Gleason s.n.!). Fig. 27C, D.

Shrub or *treelet,* 0.5–2 m tall; branchlets, petioles, and peduncles strigose with straight hairs, with globose sessile glands. *Leaves* elliptic, 5–10 × 1.5–2.5 cm, obtuse, minutely verrucose above, acute at base, 3-nerved, the lateral primaries close to the margin, lower surface on the mid-vein and lateral primary veins strigose, the lateral secondary veins, veinlets, and areoles with sessile glands in a matrix of few-branched papillae-like hairs; petioles 3–5 cm long. *Inflorescence* a many-flowered terminal paniculate cyme. *Flowers* with pedicels 3–5 mm long; pedicels, bracts, hypanthium, and calyx lobes sericeous, with sessile glands; hypanthium 1.7–2.2 × 1.7–2.2 mm; calyx lobes triangular, ca. 2 mm long; petals basally white, turning red with age, apically pink, 7–9 × 5–7 mm; filaments 7.8–8.3 or 5.7–6.2 mm long, ventrally with short-stalked glands; thecae ca. 3 mm long, the connective distinctly prolonged and expanded dorsi-basally; ovary apically sericeous with eglandular hairs, 4-locular. *Fruit* capsular. *Seeds* 1.2–1.4 mm long, subcochleate.

Distribution and ecology (Fig. 33C): *Macairea spruceana* occurs in low-lying savannas of the upper Rio Negro.

Representative specimens examined. VENEZUELA. AMAZONAS: Atabapo, N bank of lower part of Caño Caname, ca. 3 km W of Macaval, 3°41′N, 67°23′W, 95 m, *Davidse et al. 17076* (MO); Dept. Río Negro, bajo Río Siapa, 1°57′N, 66°02′W, 125 m, *Huber & Medina 5778* (B, VEN); Caño Pimichín, below Pimichín, 125–135 m, *Maguire & Wurdack 35611* (F, MO, NY, RB, US, VEN); Caño Hechimoni, 8 km above mouth of Río Siapa, 100–130 m, *Maguire et al. 37610* (F, NY, US, VEN); Maroa, 2°45′N, 67°35′W, *Steyermark & Bunting 102841* (VEN); Yavitá, 128 m, *L. Williams 14080* (F, G, NY, US, VEN).

Macairea spruceana is beautifully illustrated in Cogniaux (1885, pl. 59, fig. 1).

6. Macairea sulcata Triana, Trans. Linn. Soc. London **28:** 38. 1871. Type: Peru. Loreto: Mayo river nr. Tarapoto, Jul–Aug 1856 (fl), *Spruce 4004* (holotype: K!, photo, H. A. Gleason s.n.; isotypes: BM!, BR!, K!, W!, F photo neg. 32324!). Fig. 27A, B.

Shrub or *treelet*; branchlets, petioles, and peduncles strigulose to setulose with straight hairs, with globose sessile glands. *Leaves* elliptic or obovate, 5–11 × 2.5–4 cm, cuspidate, acute at base, 3-nerved, pseudo-pinnately veined with an inconspicuous pair of lateral primary veins arising at the blade base and close to the margin, the lateral secondary veins and the veinlets thin and densely reticulate, smooth above, lower surface on the mid-vein with few straight hairs, with sessile glands; petioles 0.6–1.2 cm long. *Inflorescence* a many-flowered terminal paniculate cyme. *Flowers* with pedicels ca. 5 mm long; pedicels, bracts, hypanthium, and calyx lobes strigulose, with sessile glands; hypanthium 1.8–2.2 × 1.8–2.3 mm; calyx lobes subulate, ca. 3 mm long; petals white or violet, 8–10 × 4–5 mm; filaments 9.8–10.20 or 6.8–7.2 mm long, ventrally with short-stalked glands; thecae ca. 4 mm long, the connective prolonged and expanded dorsi-basally; ovary apically sericeous with eglandular hairs, 4-locular. *Fruit* capsular. *Seeds* subcochleate, ca. 1 mm long.

Distribution and ecology (Fig. 33C): *Macairea sulcata* has a disjunct distribution, being known from low sandstone outcrops near Tarapoto in Peru and from sandstone terraces near Mitú on the Vaupés in Colombia.

Representative specimens examined. COLOMBIA. AMAZONAS: Araracuara, between hospital and airport, *Aguirre-Galviz 1070* (COL). VAUPES: Río Vaupés, Circasia, general loc. 0°45′N, 70°30′W, 260 m, *Schultes & Cabrera 19691* (BM, F, NY, US); Cerro de Circasia, 3–500 m, *Cuatrecasas 7197* (COL, F, NY); Río Piraparaná (trib. of Río Apaporis), Caño Paca, general loc. between 0°15′ and 0°25′N, 70°30′W, *Schultes & Cabrera 17589* (US).

There are two sheets of the type collection (*Spruce 4004*) at K, one with the locality given as "in cacumine montis Guayrapurima," the other as "in montibus flumen Mayo prope Tarapoto." Spruce (1908, vol. 2: 39) mentions collecting along a trail leading from Tarapoto eastward which crossed the "abrupt ridges of Guayrapurima ("where the wind blows") two leagues from Tarapoto" (about 9.5 km). Thus, the two indications probably refer to the same locality.

Macairea sulcata is similar to *M. spruceana* in indument and habit; it differs from that and all other species in its apically cuspidate leaves.

7. Macairea rufescens De Candolle, Prodr. **3:** 109. 1828a. Type: Colombia. Amazonas: Araracuara Mountains (fr), *Martius 3156* (holotype: M!, F photo neg. 6340!; isotype: M!). Figs. 23B, 29A–D, 34.

Shrub, 1–2 m tall; branchlets, petioles, peduncles, and mid-veins on the lower leaf surface hirsute to sericeous with long subulate hairs, with globose sessile glands. *Leaves* broadly elliptic, 7–15 × 4–7 cm, basally and apically obtuse, 3-nerved, pseudo-pinnately veined, mid-vein thick and raised, the lateral primary veins arising at the blade base, inconspicuous and close to the margin, upper surface smooth, lower surface sparsely hirsute with long subulate white hairs which are apically frayed, the areoles with sessile glands in a matrix of short, few-branched hairs; petioles 1.5–2.5 cm long. *Inflorescence* a many-flowered terminal paniculate cyme. *Flowers* with pedicels 2–3 mm long; pedicels, bracts, hypanthium, and calyx lobes sericeous with rust-colored slightly curled hairs, with sessile glands; hypanthium 3.5–4 × 1.8–2.2 mm; calyx lobes triangular, (2–)3(–4) mm long; petals basally white, turning red with age, apically pink, 9–13 × 6–7 mm; filaments 8.5–9 or 6–7 mm long, ventrally with short-stalked glands; thecae 3.5–4.5 mm long, the connective prolonged and expanded dorsi-basally; ovary apically with glandular hairs, 4-locular. *Fruit* capsular. *Seeds* ca. 0.8 mm long, subcochleate.

Distribution and ecology (Fig. 33D): *Macairea rufescens* occurs on granitic substrates (lajas) along the Río Guainía (Venezuela) and on Roraima sediment near Araracuara (Colombia); it has recently been collected near Iquitos (Peru).

Representative specimens examined. COLOMBIA. AMAZONAS: Araracuara, *Sastre 2123* (COL).

VENEZUELA. AMAZONAS: Dept. Casiquiare, Maroa, 2°43′N, 67°32′W, 100 m, *Huber 2579* (VEN); 0.5 km E of Maroa, *Maguire et al. 41709* (F, G, NY, S, US, VEN); San Carlos de Río Negro, 1–3 km on Solano rd., *Renner 2106* (US, VEN); nr. San Carlos de Río Negro, *Spruce 3134* (BM, BR, G, G-DC, K, OXF, W).

PERU. LORETO: Nr. Iquitos, *Revilla 4326* (HBG).

Macairea rufescens is distinguished from all other species by its large, leathery, pseudo-pinnately veined leaves. Such venation has arisen several times, independently, within the Melastomataceae, for example, in *Henriettella squamata* Alain (Miconieae) and in several species of *Graffenrieda* (Merianeae).

8. Macairea maroana Wurdack, Mem. New York Bot. Gard. **10:** 141. 1964. Type: Venezuela. Terr. Fed. Amazonas: Occasional at forest edges just SE of Maroa, Río Guainía, 120–140 m, 9 Oct 1957 (fl), *Maguire, Wurdack & Keith 41772* (holotype: US!; isotypes: NY!, P!, VEN!). Figs. 25B, 26G–I.

Shrub, 1.2–2 m tall; branchlets, petioles, and peduncle sericeous with long slender hairs, with globose sessile glands. *Leaves* elliptic, sometimes with denticulate margins, 7–11 × 3.5–5 cm, acuminate, obtuse at base, 3-nerved, the lateral primaries arising at the blade base, densely covered with conical hairs to 1 mm long, lower surface foveate, moderately hirsute with slender hairs, with sessile glands; petioles 1–2 cm long. *Inflorescence* a many-flowered terminal paniculate cyme. *Flowers* with pedicels 4–9 mm long; pedicels, hypanthium, and calyx lobes sericeous with glandular hairs; hypanthium 4.8–5.2 × 2.1–2.6 mm; calyx lobes narrowly triangular, ca. 3 mm long; petals white, becoming pink-flushed with

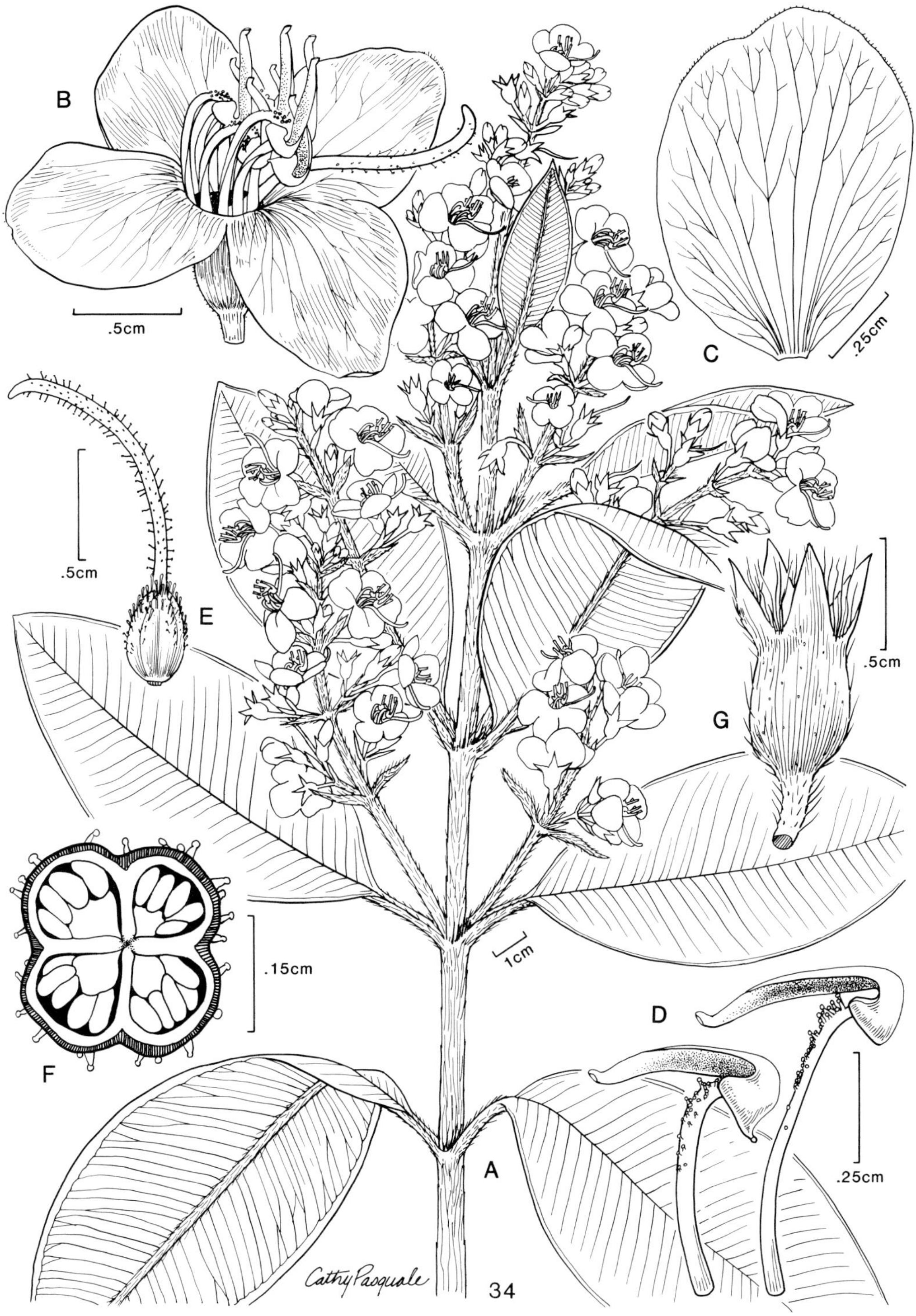

FIG. 34. *Macairea rufescens*. **A.** Habit. **B.** Flower. **C.** Petal. **D.** Stamens. **E.** Pistil. **F.** Ovary, cross-section. **G.** Young fruit.

age, 10–13 × 8–9 mm; filaments 8–9 or 5–6 mm long, ventrally with short-stalked glands; thecae ca. 5 mm long, the connective prolonged and expanded dorsi-basally; ovary apically with glandular hairs, 4-locular. *Fruit* capsular. *Seeds* 0.9–1 mm long, subcochleate.

Distribution and ecology (Fig. 35A): *Macairea maroana* is an endemic of the Orinoco-Río Negro drainage where it grows in regularly inundated savannas.

Specimens examined. VENEZUELA. AMAZONAS: Maroa, Río Guainía, 127 m, *L. Williams 14252* (F, G, NY, US).

BRAZIL. AMAZONAS: São Gabriel da Cachoeira, Alto Rio Negro, *Pires & Marinho 15709* (MG, US); Rio Içana, *Fróes 28031* (IAN).

Macairea maroana is illustrated in the *Flora de Venezuela* (Wurdack, 1973, fig. 11). It differs from all other species of *Macairea* in having long eglandular hairs and sessile glands on the lower leaf surface and long glandular hairs (but no sessile glands) on the pedicel, hypanthium, and calyx lobes.

9. Macairea cuieirasii Renner, sp. nov. Type: Brazil. Amazonas: Rio Cuieiras, igarapé da Cachoeira, campina on white sand, 15 Sep 1964, *W. Rodrigues 6039* (holotype: INPA!; isotypes: F!, US!). Fig. 25D.

Macaireae pachyphyllae Bentham affinis, floribus majoribus lamina 3-nervata nervis secundariis circa 3–6 mm inter se distantibus differt.

Shrub or *treelet,* 2–4 m tall; branchlets, petioles, and peduncles sericeous with slightly curled, basally broad, glandular hairs, the gland heads caducous. *Leaves* narrowly obovate, (4.5–)7–9 × (1.8–)3–3.5 cm, apically acute or obtuse, acute at base, 3-nerved, the lateral primaries arising at the blade base and close to the margin, the lateral secondary veins 3–6 mm apart, upper surface densely covered with conical hairs to 1 mm long, lower surface foveate, the mid-vein, lateral primary veins, lateral secondary veins, veinlets, and areoles hirsute with short, weak, glandular hairs; petioles 0.8–1.8 cm long. *Inflorescence* a many-flowered terminal paniculate cyme. *Flower* with pedicels 2–4 mm long; pedicels, hypanthium, and calyx lobes hirsute with glandular hairs; hypanthium 2.8–3.2 × 3.1–3.7 mm; calyx lobes triangular, 2–3 mm long; petals basally white, turning red with age, apically pink, 7–9 × 4–6 mm; filaments 8.1–8.7 or 6.1–6.7 mm long; ventrally with short-stalked glands; thecae ca. 4 mm long, the connective distinctly prolonged and expanded dorsi-basally; ovary apically with few glandular hairs, 4-locular. *Fruit* capsular. *Seeds* ca. 0.9 mm long, subcochleate.

Distribution and ecology (Fig. 35C): *Macairea cuieirasii* occurs in white-sand savannas (called "campinas" in Brazilian Amazonia) along the lower Rios Negro and Branco (Amazônia, Brazil).

Additional specimens examined. BRAZIL. AMAZONAS: Rio Univini, *Pires et al. 14215* (IAN, MG, US); Rio Cuieiras, Igarapé da Cachoeira, *Rodrigues & Wilson 3978* (INPA, US), *Rodrigues & Coêhlo 4874* (INPA, US).

Macairea cuieirasii is similar to *M. pachyphylla,* but it has larger flowers, obovate leaves, and more widely separated lateral secondary veins. The second species is widespread, occurring from the Gran Sabana in Venezuela, through Guyana and Surinam, south to Bolivia; *Macairea cuieirasii,* on the other hand, seems to be a relatively narrow endemic, geographically isolated from its putative closest relative.

10. Macairea pachyphylla Bentham, J. Bot. (Hooker) **2:** 292. 1840. Type: Guyana. Swampy savannas nr. Mt. Roraima (Wawemiddimie), 1837 (fl), *Ro. Schomburgk 452* (holotype: K!; isotypes: BM!, BR frag.!, F!, G!, G-DC!, NY!, OXF!, P!, U!, W!; B destroyed, but represented by a photo, F neg. 16701!). Fig. 27E, F.

Macairea calvescens Naudin, Ann. Sci. Nat. Bot. III, **13:** 35. 1850a. Type: Guyana. Mt. Roraima, 1842/43 (fl), *Ri. Schomburgk 521* (holotype: P!; isotypes: BM!, G!, K!, NY!, OXF!, W!).

Macairea foveolata Cogniaux *in* Martius, Fl. bras. **14(4):** 596. 1888. Type: Brazil. Amazonas: Alto Amazonas (fl), *Glaziou 9806* (prob. Schwacke leg.) (holotype: BR!; isotypes: C!, F photo neg. 21181!, P!).

Macairea hoehnei Cogniaux, Comm. Linh. Telegr. Estr. Matto Grosso, Annexo 5, Bot. **3:** 6. 1912. Type: Brazil. Mato Grosso: Utiariti, Jun 1909 (fl), *Hoehne 2051* (lectotype, here designated: BR!; isolectotype: US frag.!, US photo s.n.!).

Macairea viscosa Ducke, Arq. Jard. Bot. Rio de Janeiro **3:** 223. 1922. Type: Brazil. Pará: Serra de Parauaquara, nr. Praínha, 250–300 m, 7 Oct 1919 (fl), *Ducke s.n.* [RB 2398] (holotype: RB!; isotypes: G!, IAN, K!, S!; B destroyed, but represented by a photo, F neg. 16706!).

Macairea cordeiroi Brade, Arq. Jard. Bot. Rio de Janeiro **16:** 8, fig. 3. 1959a. Type: Brazil. Rondônia:

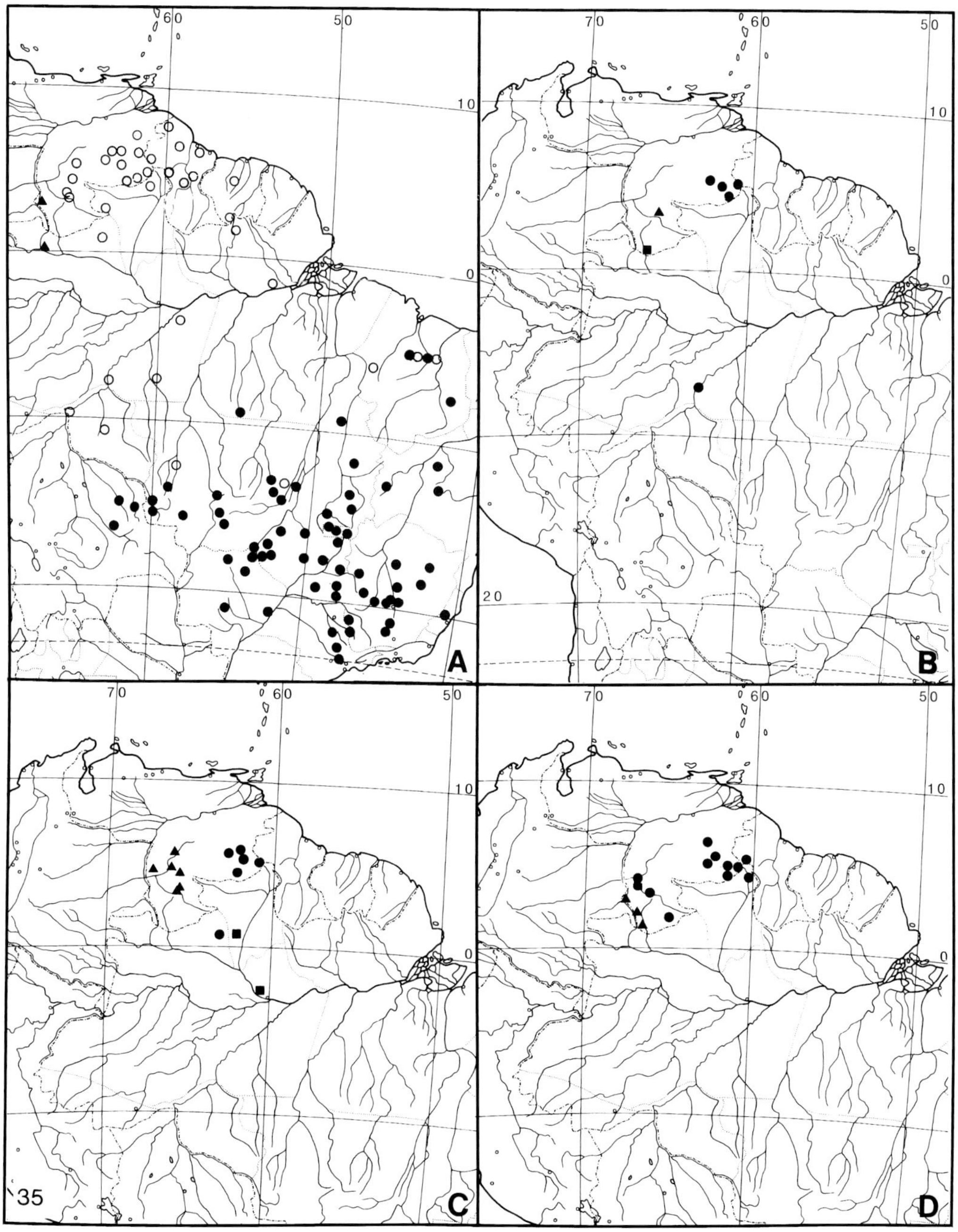

FIG. 35. Distribution of *Macairea*. **A.** *M. radula* (●), *M. pachyphylla* (○), *M. maroana* (▲). **B.** *M. multinervia* (●), *M. linearis* (▲), *M. neblinae* (■). **C.** *M. parvifolia* (●), *M. rigida* (▲), *M. cuieirasii* (■). **D.** *M. lasiophylla* (●), *M. axilliflora* (▲).

Porto Velho, 1952 (fl), *Cordeiro & Silva 311* (holotype: RB!; isotype: NY!).

Shrub or *treelet,* 1.5–4(–12) m tall; branchlets, petioles, and peduncles loosely hirsute with glandular hairs, hairs to 7 mm long. *Leaves* elliptic, 6–15 × 3–7(–9) cm, basally and apically obtuse or acute, 5-nerved, the lateral primaries arising at the blade base, the outermost pair close to the margin, the lateral secondary veins 2–4 mm apart, upper surface densely hirsute with patent, conical, glandular hairs to 1 mm long, lower surface foveate, the mid-vein, lateral primary veins, and lateral secondary veins with subulate glandular hairs, the areoles with minute, weak, glandular hairs; petioles 1–2 cm long. *Inflorescence* a many-flowered terminal paniculate cyme. *Flowers* with pedicels 2–4 mm long; pedicels, bracts, hypanthium, and calyx lobes loosely hirsute to sericeous with glandular hairs, the gland heads caducous; hypanthium 1.5–2 × 1.8–2.2 mm; calyx lobes narrowly triangular, 2–3(–6) mm long; petals basally white, turning pink with age, apically pink, 7–8.5 × 5–6 mm; filaments (3–)5–7.5 or 3–4 mm long, ventrally with short-stalked glands; thecae 2(–3) mm long, the connective minutely prolonged dorsi-basally; ovary apically with glandular hairs, 4-locular. *Fruit* capsular. *Seeds* 0.6–0.7 mm long, subcochleate.

Distribution and ecology (Fig. 35A): *Macairea pachyphylla* occurs from the Gran Sabana in Venezuela, through Guyana, Surinam, and Brazilian Amazonia, south into Bolivia; from sea level to 1250 m alt.

Representative specimens examined. VENEZUELA. AMAZONAS: Dept. Atabapo, Serranía del Parú, 4°32′N, 65°31′W, 800 m, *Huber 3616* (VEN); Rio Cunucunuma medio, pie SE de Cerro Huachamacari, 3°48′N, 65°43′W, 400 m, *Huber 4984* (VEN); Cerro Duida, 03°10′N, 65°31′W, 1200 m, *Tillett et al. 752-305* (NY, US). BOLIVAR: Summit of Cerro Muru, 1500 m, *Cardona 1840* (G,VEN); Orillas Quebrada, Auyántepuí, Alto Caroní, 1100 m, *Cardona 2617* (NY, S, US, VEN); Distr. Piar, top of Salto Aicha nr. E base of Uaipán-tepuí, 5°38′N, 62°32′W, 1100 m, *Davidse & Huber 22901* (MO, US); 17 km E of El Pauji by rd. and 64 km W of Sta. Elena by rd., 4 km N of hwy., Río Las Ahallas, ca. 4°30′N, 61°30′W, 850 m, *Liesner 19099* (MO); ca. 4 km E of Mission Sta. Teresita de Kavanayen, 1300 m, *Maguire & Wurdack 33989* (F, MO, NY, US, VEN).

GUYANA. Georgetown, 2 mi E of Atkinson Field (Georgetown airport), *Irwin BG-18* (US); Pakaraima Mts., Mt. Aymatoi, ca. 5°55′N, 61°W, 1150 m, *Maas et al. 5823* (NY, U, US); Kaieteur plateau, *Maguire & Fanshawe 23342* (NY, US); upper Mazaruni river basin, nr. mouth of Mokay river, 490 m, *Tillett & Tillett 45348* (NY, US, VEN).

SURINAM. Tafelberg, 1.5 km S of savanna I, *Maguire 24350* (F, MO, NY, RB, US); Sipaliwini savanna area on Brazilian frontier, *Oldenburger et al. 1418* (NY, U, US).

BRAZIL. AMAZONAS: Mun. Barcelos, plato da Serra Aracá, parte SE da Serra Norte, 0°1′N, 63°22′W, 1150–1250 m, *Alencar 1626* (INPA, NY, US); Rio Branco, slopes of Serra Maturacá, *Rodrigues 90* (INPA). MARANHÃO: Ca. 40 km W of Caxías on BR 135 near km 512, ca. 4°42′S, 43°47′W, *Daly et al. 707* (MG, NY, US); Burití, *Schwacke 566* (RB). MATO GROSSO: Ca. 3 km W of base camp, 12°49′S, 51°46′W, *Harley & Souza 10206* (P, RB, US). PARÁ: Praínha, Serra de Parauquara, *Ducke s.n.* [RB 2398] (G, K, P, RB, S); Tapajóz, Rio Cururú, *Egler 808* (IAN, US). RONDÔNIA: Km 220–223 Madiera–Marmoré railrd. nr. Abunã, *Prance et al. 5983* (MG, MO, NY, S, US, VEN); Pacas Novos, 10°46′S, 63°49′W, *Rosa et al. 881* (MG, MO, NY). RORAIMA: Serra dos Surucucus, 2°42–47′N, 63°33–36′W, *Prance et al. 9900* (F, NY, S, US); Rio Branco, Surumú, Serra de Mairarí, *Ule 8442* (G, K, U).

BOLIVIA. BENI: San Josa (José?), 600 m, *R. S. Williams 440* (NY, US).

Macairea pachyphylla is apparently well adapted to growing in seasonally disturbed habitats such as gallery forest and flooded savannas; it has also made its way into burned forest land and strongly degraded vegetation (e.g., Atkinson airfield near Georgetown).

One of the characteristic features of this species is the loose pubescence consisting of rather long, patent, glandular hairs present on the entire plant. In related species (*Macairea rigida, M. parvifolia, M. neblinae,* and *M. radula*), the pubescence is shorter and, particularly on the leaves, much denser. Another close relative, *M. cuieirasii,* differs in leaf shape and venation (compare discussion under *M. cuieirasii*). A good drawing of *M. pachyphylla* is provided by Brade (1959a, as *M. cordeiroi*).

11. Macairea radula (Bonpland) De Candolle, Prodr. **3:** 109. 1828a. Fig. 28D–F.

Rhexia radula Bonpland *in* Humboldt & Bonpland, Monogr. Melast. **2:** 107, tab. 41. 1820. Type: Brazil. (fl), *Ferreira s.n.* (*=223*), 1790/91 (holotype: P!, F photo neg. 36142!; isotypes: LISU!; B destroyed, but represented by a photo, H. A. Gleason s.n.!).

Macairea adenostemon De Candolle, Prodr. **3:** 109, 484. 1828a. Type: Brazil. Mato Grosso: Nr. Cuiabá, Jun 1833 (fl), *Manso 177* (*=Martius Herb. Fl. 210*) (holotype: G-DC!; isotypes: BR!, G!, K!, LE!, M!, MO!, NY!, P!, W!).

Macairea adenostemon var. *ursina* Schrank & Martius ex De Candolle, Prodr. **3:** 109. 1828a. Type: Brazil.

Minas Geráis: Sequetinhonha river (fl), *Martius s.n.* (holotype: M!, F photo neg. 6338!).

Macairea adenostemon var. *martiana* Cogniaux *in* Martius, Fl. bras. **14(3):** 245. 1885 & **14(4):** 569 (addenda). 1888. This includes the type of the name *Macairea adenostemon* De Candolle and should be replaced by *Macairea adenostemon* DC. var. *adenostemon.*

Macairea adenostemon var. *rotundata* Pilger, Bot. Jahrb. Syst. **30:** 179. 1902 (1901). Type: Brazil. Mato Grosso: Upper Rio Ronuro, infrequent in dry campo at forest edge, 25 May 1899 (fl), *Meyer 634* = *Pilger 634* (holotype at B destroyed, but represented by a photo, H. A. Gleason s.n.).

Macairea mosenii Cogniaux *in* Martius, Fl. bras. **14(3):** 242, pl. 57. 1885 & **14(4):** 595 (addenda). 1888. Type: Brazil. São Paulo: Between Casabranca and Cajurú, Mar 1857 (fl), *Regnell III.n.20* (lectotype, here designated: S!; isolectotypes: C!, S!, US!).

Macairea sericea Cogniaux *in* Martius, Fl. bras. **14(3):** 243, tab. 58. 1885. Type: Brazil. Minas Geráis: Nr. João d'El Rei, s.a. (fl), *Pohl 620* (=*1228*) (lectotype, here designated: BR!; isolectotypes: NY!, W!).

Macairea ledifolia Cogniaux *in* Martius, Fl. bras. **14(3):** 245. 1885. Type: Brazil. Bahía: S. loc. [prob. nr. Salvador along Paraguaçú river], Nov 1818 (fl), *Martius s.n.* (holotype: M!, F photo neg. 6339!).

Macairea rotundifolia Cogniaux, Comm. Linh. Telegr. Estr. Matto Grosso, Publ. 10, Annexo 5, Bot. **3:** 7. 1912. Type: Brazil. Mato Grosso: Tres Jacús, Jun 1909 (fl), *Hoehne 2184* (holotype: BR frag.!).

Macairea goyazensis Hoehne, Mem. Inst. Butantan, Sec. Bot. **1(5):** 60, pl. 8, fig. 2. 1922. Type: Brazil. Goiás: São Gonçalo, 1912 (fl), *Lützelburg 1280* (holotype: RB!).

Macairea moraisiana Brade, Arq. Jard. Bot. Rio de Janeiro **16:** 30, pl. 3, figs. 1–10. 1959b. Type: Brazil. Goiás: Niquelândia, 26 Feb 1956 (fl), *Macedo 4459* (holotype, HB!).

Macairea macedoi Brade, Arq. Jard. Bot. Rio de Janeiro **16:** 31, pl. 3, figs. 11–17. 1959b. Type: Brazil. Goiás: Jataí, Ribeirão Grande, in várzea vegetation, 1 Aug 1956 (fl), *Macedo 4631* (holotype: HB!; isotypes: S!, SP!, US!).

Shrub, 1.5–2 m tall; branchlets, petioles, and peduncles sericeous with slightly curled glandular hairs. *Leaves* broadly elliptic to narrowly elliptic, 5–8 × 2.5–4 cm, obtuse or minutely apiculate, acute at base, 5-nerved, the lateral primary veins arising at the blade base, the outermost pair close to the margin, upper surface densely hirsute with conical glandular hairs, gland heads caducous, lower surface densely sericeous with long, thin, slightly curled glandular hairs, the lateral secondary veins outlined with dense pubescence of long, slightly curled hairs, the areoles hirsute with shorter glandular hairs; petioles (0.5–)1–2 cm long. *Inflorescence* a many-flowered terminal paniculate cyme. *Flowers* with pedicels (1–)2–3 mm long; pedicels, bracts, hypanthium, and calyx lobes hirsute with glandular hairs; hypanthium 1.8–2.2 × 1.7–2.2 mm; calyx lobes subulate, 2–3 mm long; petals basally white, turning red with age, apically pink, 7–8 × 4–5 mm; filaments 10–11 or 8–9 mm long, ventrally with short-stalked glands; thecae 4–5 mm long, the connective minutely prolonged and expanded dorsi-basally; ovary apically with glandular hairs, 4-locular. *Fruit* capsular. *Seeds* 0.5–0.7 mm long, subcochleate.

Distribution and ecology (Fig. 35A): *Macairea radula* is widespread in central Brazil where it grows in cerrado vegetation ("campo rupestre"), grassy savannas, and gallery forests at elevations from a few hundred meters up to 950 m.

Representative specimens examined. BRAZIL. BAHIA: Serra de Acuruá, *Blanchet 2900* (BM, G, G-DC, K, NY, OXF, P, W); Mucurí, *Sellow 12* (BR, GOET). DISTRITO FEDERAL: Parque Mun. do Gama, ca. 20 km S of Brasília, *Irwin & Soderstrom 5818* (F, K, NY, R, RB, UB, US). ESPIRITO SANTO: Between Victoria and Bahia, *Sellow 130* (BR, G, US). GOIÁS: Natividade, *Gardner 3162* (BM, K, NY, OXF, W); Serra do Caiapó, ca. 40 km S of Caiapônia, 900 m, *Irwin et al. 17769* (MO, NY, R, S, SP, U, UB, US); Serra dos Pireneus, ca. 18 km E of Pirenópolis, 1000 m, *Irwin et al. 34251* (F, NY, R, UB, US); Rio Beason, Natividade, *Pohl 1260* (BM, NY, W). MARANHÃO: Barra do Corda, 200 m, *Lisbôa 2344* (B, BM, G, RB, US); between Theresina and Caxías, *Jobert 1232* (P); Alto Parnaiba, Victoria, *Snethlage 702* (F). MATO GROSSO: Col. Paxixi, Mun. Aquidauana, *Hatschbach & Guimarães 24593* (C, COL, HBG, MO, NA, NY, RB, S, US); ca. 20 km S of Xavantina, 400 m, *Irwin et al. 16818* (F, MO, NY, P, R, RB, S, UB, US); Sararé, 58°55′W, 13°50′S, *Pires & Santos 16405* (F, NY, RB, US). MINAS GERAIS: Patos de Minas, 800 m, *Duarte 2814* (BM, BR, NY, P, U, US); Itabira, *Glaziou 16787* (BR, C, G, K, LE, P, US); Serra do Espinhaço, Mun. Gouveia, 1200 m, *Hatschbach 27334* (C, NY, S, US); Sorocaba, *Lund s.n.* (C, LE, S, US); Mun. Nova Lima, Acaba Mundo, Serra do Curral, 1000 m, *Williams & Assis 8066* (BM, BR, C, COL, F, G, LE, MO, RB, S, U, US, YU). PARÁ: Mun. de Itaituba, estr. Santarém-Cuiabá, BR 163, km 794, Serra do Cachimbo, Aeroporto do Cachimbo com penetração de 8 km dentro da mata, *I. L. Amaral et al. 1014* (INPA, NY, US). PIAUÍ: Paranagoa, *Gardner 2587* (BM, G, G-DC, K, NY, OXF, P, US, W). RIO DE JANEIRO: Nr. Rio de Janeiro, *Glaziou 13843* (K, P); Mello Verde, *Saint-Hilaire s.n.* (P). SÃO PAULO: Moji-Guacú, ca. 4 km NNW of Padua Sales, *Eiten & Eiten 1743* (F, NY, SP, US); Campinas, *Heiner s.n.* (S); Monte Alegre, Araraquara, *Löfgren 1145* (BR frag., C, P, SP); between Casabranca and Cajurú, *Regnell III/20* (C, S, US); Sorocaba, *Riedel 2136* (BR, G, GOET, LE, MO, NY, OXF, S, US); São Paulo, *Sellow s.n.* (LE).

BOLIVIA. SANTA CRUZ: Nr. the Arches, 5 km N of Santiago de Chiquitos, 700 m, *Cutler 7031* (NY); Chiquitos, *Orbigny 687* (BR, G-DC, P, W); Serranía de Ricardo Franco, *Schmidt 43* (HBG, M).

Macairea radula, in spite of the many named varieties, is readily recognized by the dense pubescence of the leaves, consisting of rather long hairs, which on the lower surface give the lateral secondary veins a wavy appearance.

Macairea radula is illustrated in Bonpland (1820), de Candolle (1828b), Cogniaux (1885, pl. 57, as *M. mosenii* and pl. 58, as *M. sericea*), and Brade (1959b).

12. Macairea parvifolia Bentham, J. Bot. (Hooker) **2:** 293. 1840. Type: "Guyana." Mt. Roraima, Dec 1842, *Ri. Schomburgk 1089B* (=*699*) (holotype: K!; isotypes: BM!, BR!, G!, K!, NY!, P!, US!, W!; B destroyed, but represented by a photo, H. A. Gleason s.n.!).
Fig. 36.

Shrub, 0.8–2 m tall; branchlets, petioles, and peduncles hirsute with glandular hairs, hairs to 1 mm long. *Leaves* elliptic, 2–4 × 0.8–1.8 cm, basally and apically acute or obtuse, 5-nerved, the lateral primaries arising at the blade base, the outermost pair close to the margin and inconspicuous, upper surface densely hirsute with conical glandular hairs, gland heads caducous, lower surface on the mid-vein and lateral primary veins strigose with mostly eglandular hairs with broad bases, the lateral secondary veins, veinlets, and areoles hirsute with short glandular hairs; petioles 0.5–1 cm long. *Inflorescence* a many-flowered terminal paniculate cyme. *Flowers* with pedicels 2–3 mm long; pedicels, bracts, hypanthium, and calyx lobes hirsute with glandular hairs; hypanthium 2.5–3 × ca. 2 mm; calyx lobes subulate, 2.5–3(–3.5) mm long; petals basally white, apically pink, the basal section turning red with age, 7.5–9 × 5–6 mm; filaments 6.8–7.2 or 3.7–4.2 mm long, ventrally with short-stalked glands; thecae 3–3.5 mm long, the connective dorsi-basally minutely prolonged and expanded; ovary apically with glandular hairs, 3-locular. *Fruit* capsular. *Seeds* ca. 0.7 mm long, subcochleate.

Distribution and ecology (Fig. 35C): *Macairea parvifolia* occurs on sandstone mountains of the Gran Sabana (Venezuela) at elevations of (750–)1200–1800 m.

Representative specimens examined. VENEZUELA. BOLIVAR: Río Carún, Alto Paragua, *Cardona 2468* (VEN); Distr. Roscio, N del Kamámeru, 5°25′N, 61°13′W, 1000 m, *Huber et al. 7274* (US); 17 km E of El Pauji by rd. and 64 km W of Sta. Elena by rd., 4 km N of hwy., Río Las Ahallas, 4°30′N, 61°30′W, 850 m, *Liesner 19307* (MO, US); ca. 4 km E of Mission Sta. Teresita de Kavanayén, 1300 m, *Maguire & Wurdack 33792* (F, MO, NY, RB, US, VEN); Río Karuai, between Sta. Teresita de Kavanayén and base of Ptari-tepuí, 1220 m, *Steyermark 60322* (F, NY, VEN); cumbre del Cerro Guaiquinima, 5°38′N, 63°45′W, 1650 m, *Steyermark et al. 117461* (MO, VEN); Auyán-tepuí, *Vareschi 4685* (NY, VEN).

BRAZIL. AMAZONAS: Plateau of N massif of Serra Aracá, 0°51–57′N, 63°21–22′W, 1200–1400 m, *Prance et al. 29167A* (INPA, NY).

Richard Schomburgk (1848, vol. **2:** 171, 212) mentions *Macairea parvifolia* and *M. multinervia* (treated below) as among the main representatives of the flora of Roraima and as occurring on the Humirida range, a sandstone area just southwest of Roraima toward the Venezuelan/Brazilian border. The types of these two species may therefore come from Venezuela and not from Guyana as indicated on their labels.

13. Macairea rigida Bentham, J. Bot. (Hooker) **2:** 292. 1840. Type: Venezuela. Terr. Fed. Amazonas: Marahuaca, 22–24 Feb 1839, *Schomburgk 1015* (holotype: K!, photo, H. A. Gleason neg. 89-5!; isotypes: BM!, BR frag.!, G-DEL!, P!, W!; B destroyed, but represented by a photo, F neg. 10703!).
Figs. 24D, 28A–C.

Shrub or *treelet,* 1–5 m tall; branchlets, petioles, and peduncles hirsute with glandular hairs, hairs to 1 mm long. *Leaves* obovate, 4–7 × 2–4 cm, rounded or minutely apiculate, obtuse at base, 5-nerved, the lateral primaries arising at the blade base, the outermost pair close to the margin and inconspicuous, upper surface densely covered with stout, conical, glandular hairs, lower surface foveolate, the mid-vein and lateral primary veins hirsute with basally broad, apically slightly curled glandular hairs, the lateral secondary veins and veinlets with shorter hairs of the same type; petioles 0.6–1 cm long. *Inflorescence* a many-flowered terminal paniculate cyme. *Flowers* with pedicels 3–6 mm long; pedicels, bracts, hypanthium, and calyx lobes sparsely hirsute with thin glandular hairs; hypanthium 2.5–3 × 1.8–2.2 mm; calyx lobes subulate, 2.5–3 mm long; petals basally white, turning red with age, apically pink, 6–7 × 4–5.5 mm; filaments 5.8–6.2 or 3.8–4.2 mm long, ventrally with short-stalked glands;

FIG. 36. *Macairea parvifolia.* **A**, **B.** Habits. **C.** Upper leaf surface. **D.** Lower leaf surface. **E.** Flower. **F.** Fruit. **G.** Ovary. **H.** Stamens.

thecae 2–2.5 mm long, the connective minutely prolonged and expanded dorsi-basally; ovary apically with glandular hairs, 3-locular. *Fruit* capsular. *Seeds* 0.6–0.7 mm long, subcochleate.

Distribution and ecology (Figs. 30, 35C): *Macairea rigida* occurs on the summits of Mt. Duida, Cerro Marahuaca, Serranía Parú, and Serranía Yutaje in Territorio Amazonas in Venezuela; from 800 to 2000 m elevation.

Representative specimens examined. VENEZUELA. AMAZONAS: Dept. Atabapo, Serranía Parú (Aroko), sector Norte, planicie central, 4°32′N, 65°31′W, 800 m, *Huber 3599* (US, VEN); Mt. Duida, Cerro Culebra, 1600 m, *B. & B. Maguire 29105* (NY); Mt. Duida, *Tate 769* (NY); Cerro Marahuaca, 1500 m, *B. & B. Maguire 29166* (NY); Serranía Yutaje, Río Manapiare, 1300 m, *B. & C. K. Maguire 35068* (F, MO, NY, RB, US, VEN).

14. Macairea neblinae Wurdack, Mem. New York Bot. Gard. **10:** 100. 1958b and loc. cit. **10:** 141. 1964. Type: Venezuela. Terr. Fed. Amazonas: Cerro de la Neblina, Río Yatua. Frequent on upper escarpment slopes E of camp 3, 1900 m, 24 Dec 1953 (fr), *Maguire, Wurdack & Bunting 36935* (holotype: NY!, US photo s.n.!; isotype: US!). Fig. 28G–I.

Shrub, 1–2 m tall; branchlets, petioles, and peduncles hirsute with glandular hairs. *Leaves* narrowly elliptic, (3.5–)6–8 × 2–3.5 cm, apically acute, rounded or obtuse at base, 5-nerved, the lateral primaries arising at the blade base, the outermost pair close to the margin, upper surface densely hirsute with conical glandular hairs to 2 mm long, lower surface foveolate, the mid-vein and lateral primary veins densely hirsute with basally broad, apically slightly curled glandular hairs, the lateral secondary veins and veinlets with shorter hairs of the same type, petioles 1–1.5 cm long. *Inflorescence* a many-flowered terminal paniculate cyme. *Flowers* with pedicels 2–4 mm long; pedicels, bracts, hypanthium, and calyx lobes hirsute with slightly curled glandular hairs; hypanthium 2.1–2.6 × 2.1–2.5 mm; calyx lobes subulate, to 2.5 mm long; petals basally white, turning red with age, apically pink, 6–9 × 4–6 mm; filaments 4–6 or 3.1–3.6 mm long, ventrally with short-stalked glands; thecae 2.9–3 mm long, the connective prolonged and expanded dorsi-basally; ovary apically with glandular hairs, 4-locular. *Fruit* capsular. *Seeds* ca. 0.6 mm long, subcochleate.

Distribution and ecology (Fig. 35B): *Macairea neblinae* is endemic to Cerro de la Neblina (Terr. Fed. Amazonas, Venezuela), where it grows on shallow soil on rock outcrops at elevations of 1700–2200 m.

Representative specimens examined. VENEZUELA. AMAZONAS: Cerro de la Neblina, NE of Cumbre Camp, 1850–2200 m, *Maguire et al. 42095* (B, F, G, NY, P, RB, US, VEN); 2000–2100 m, *Maguire et al. 42423* (F, G, NY, P, RB, S, US, VEN); camp II, 2.8 km NE of Pico Phelps, 00°49′N, 65°59′W, 2085–2100 m, *Renner 2008* (B, US, VEN); 1900 m, *Steyermark 103933* (COL, F, NY, US, VEN).

15. Macairea linearis Gleason, Bull. Torrey Bot. Club **58:** 417, pl. 35, fig. 1. 1931b. Type: Venezuela. Terr. Fed. Amazonas: Summit of Mt. Duida, in the stream-bed at Central Camp, 1600 m, 20 Dec 1928 (fl), *Tate 532* (holotype: NY!, NY photo neg. s.n.!; isotype: US!). Fig. 27H, I.

Shrub, to 1 m tall; branchlets lepidote with appressed lanceate hairs. *Leaves* linear, 1.2–1.8 × ca. 0.2 cm, acute, cuneate at base, 3-nerved, the pair of lateral primary veins arising at the blade base, glabrous, upper surface verrucose, lower surface foveate; petioles ca. 0.3 cm long, lepidote like the stem. *Inflorescence* a terminal 3–5-flowered paniculate cyme. *Flowers* with pedicels 4.5–5.2 mm long, lepidote; hypanthium 2.1–2.6 × 2.2–2.5 mm, inconspicuously 8-ribbed, basally and on the ribs sublepidote; calyx lobes linear-lanceolate, ca. 3.5 mm long; petals magenta, 7–9.5 × 7–8 mm; filaments 3.8–4.2 or 2.7–3.4 mm long, ventrally with short-stalked glands; thecae ca. 2.5 mm long, the connective prolonged and expanded dorsi-basally; ovary apically with glandular hairs, 3-locular. *Fruit* capsular. *Seeds* ca. 0.6 mm long, subcochleate, testa isodiametrically foveate.

Distribution and ecology (Fig. 35B): *Macairea linearis* is endemic to Mt. Duida where it grows along stream beds at 1600–1980 m.

Additional specimen examined. VENEZUELA. AMAZONAS: Summit of Cerro Duida, 1700–1980 m, *Steyermark 58142* (F, NY).

Mt. Duida is rich in plant species with linear leaves; besides *Macairea linearis,* there are *Clidemia linearis* (Gleason) Wurdack, *Adelobotrys stenophylla* Wurdack, and *Poteranthera duidae* (Gleason) Wurdack in the Melastomataceae, *Tyleria linearis* Gleason (Ochnaceae), *Ternstroemia duidae* Gleason (Theaceae), *Ravenia linearis* Gleason (Rutaceae), and *Duidaea rubriceps* Blake, *D. tatei* Blake, *D. pinifolia* Blake, and *Tyleropappus dichotomus* Greenm. in the Asteraceae.

A photograph of the type specimen of *Macairea neblinae* accompanies the protologue.

16. Macairea cardonae Wurdack, Mem. New York Bot. Gard. **10:** 139, fig. 64a–d. 1964. Type: Venezuela. Bolívar: Chimantá Massif, cumbre of Acopán-tepuí, 2200 m, Oct 1947 (fl), *Cardona 2238* (holotype: US!).
Fig. 25A, C.

Shrub, 0.5–2 m tall; branchlets, petioles, and peduncles hirsute with glandular hairs to 2 mm long. *Leaves* ovate, 3.5–7 × 3–4.5 cm, obtuse to acute, rounded at base, 5–7-nerved, the lateral primaries arising at the blade base, the outermost pair inconspicuous and close to the margin, upper surface densely hirsute with conical glandular hairs to 4 mm long, lower surface densely hirsute on the mid-vein, lateral primary veins, lateral secondary veins, and veinlets, with slightly curled glandular hairs; petioles 0.5–1.5 cm long. *Inflorescence* a terminal, congested, few-flowered paniculate cyme. *Flowers* with pedicels 3–5 mm long; pedicels, bracts, hypanthium, and calyx lobes hirsute with glandular hairs; hypanthium 2.8–3.2 × 2.8–3.1 mm; calyx lobes triangular, ca. 3 mm long; petals basally white, turning red with age, apically deep orchid rose, 12–15 × 10–15 mm; filaments 5–6 or 4–5 mm long, ventrally with short-stalked glands; thecae ca. 4 mm long, the connective minutely prolonged and expanded dorsi-basally; ovary apically with glandular hairs, 4-locular. *Fruit* capsular. *Seeds* 1–1.1 mm long, subcochleate.

Distribution and ecology (Fig. 37A): *Macairea cardonae* is common on the Chimantá Massif and Auyán-tepuí (Estado Bolívar, Venezuela) at 1940–2500 m elevation.

Representative specimens examined. VENEZUELA. BOLIVAR: Afl. del Icabarú, Caroní, *Cardona 1866* (NY); Auyán-tepuí, *Cardona 256* (US, VEN); Distr. Piar, Chimatá Massif, 5°18′N, 62°03′W, 2450 m, *Huber & Steyermark 7156* (US, VEN); Chimantá, sector SSE, Acopán-tepuí, cabeceras del Río Arauác, 1920 m, *Luteyn et al. 9524* (NY, US); Chimantá, upper shoulder of Apácara-tepuí, *Steyermark 75844* (F, NY); plateau of Auyán-tepuí, 2300 m, *Vareschi 4867* (NY); Churí-tepuí (Muru-tepuí) of Chimantá Massif, 2250–2300 m, *Wurdack 34220* (NY, US).

A drawing accompanies the protologue of *Macairea cardonae,* and there is a close-up photograph of the leaves in Brewer Carías (1978: 76).

17. Macairea lanata Gleason, Bull. Torrey Bot. Club **58:** 416. 1931b. Type: Venezuela. Terr. Fed. Amazonas: Summit of Mt. Duida, Brocchinia Hills, 1500 m, 4 Jan 1929, *Tate 588* (holotype: NY!, NY photo s.n.!).
Figs. 24E, 26A–C.

Macairea lanata ssp. *eglandulosa* Wurdack, Mem. New York Bot. Gard. **10:** 99. 1958b. Type: Venezuela. Terr. Fed. Amazonas: Abundant in sabanita 1 km east of Maroa, Río Guainía, 120–140 m, 27 Nov 1953 (fl), *Maguire, Wurdack & Bunting 36440* (holotype: NY!, NY photo neg. 5644!; isotypes: S!, US!).

Shrub, 0.2–1 m tall; branchlets, petioles, and peduncles hirsute to subsericeous with glandular hairs (gland heads caducous) to 10 mm long. *Leaves* elliptic or ovate, 2.5–7 × 1.5–2.8 cm, obtuse, rounded at base, 5-nerved, the lateral primaries arising at the blade base, upper surface densely covered with glandular hairs 1–4 mm long, the gland heads caducous, lower surface foveate, the mid-vein and lateral primary veins hirsute with glandular hairs to 6 mm long (gland heads caducous); petioles 0.1–0.6 cm long. *Inflorescence* consisting of few-flowered paniculate cymes in the axils of the uppermost leaves. *Flowers* with pedicels 4–8 mm long; pedicels, bracts, hypanthium, and calyx lobes hirsute with slightly curled glandular hairs; hypanthium 2.8–3.2 × 1.8–2.1 mm; calyx lobes triangular, 2–3 mm long; petals pink to magenta, 10–11 × 5–7 mm; filaments 5.5–9.5 or 4–5 mm long, ventrally with short-stalked glands or glabrous; thecae 3–4 mm long, the connective prolonged and expanded dorsi-basally; ovary apically with glandular hairs, 3(–4)-locular. *Fruit* capsular. *Seeds* ca. 1 mm long, subcochleate to polyhedric.

Distribution and ecology (Fig. 37A): *Macairea lanata* is known from the summits of the sandstone mountains Parú, Moriche, Duida, and Aracamuni and from swampy lowland savannas of Amazonian Venezuela and Colombia; 100–1400 m elevation.

Representative specimens examined. COLOMBIA. GUAINIA: 0.5–1.5 km N of Puerto Colombia, opposite Maroa, 130 m, *Maguire et al. 41847* (COL, F, NY, S, US); Correg. de Puerto Colombia, Río Negro, 75 m, *Pabón et al. 423* (COL).

VENEZUELA. AMAZONAS: Sta. Cruz, margin of Río

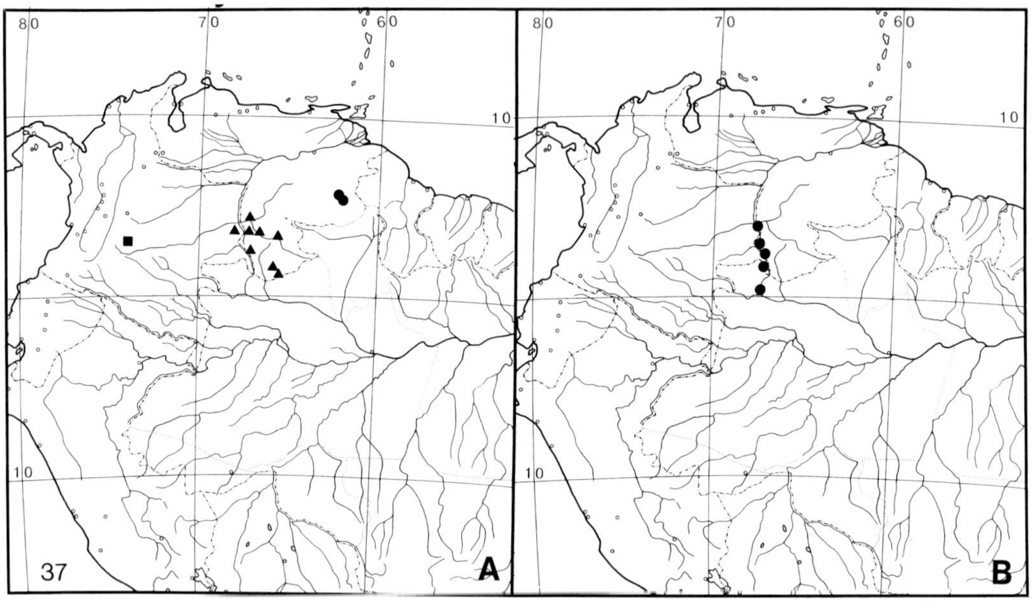

FIG. 37. Distribution of *Macairea*. **A.** *M. cardonae* (●), *M. lanata* (▲), *M. philipsonii* (■). **B.** *M. stylosa*.

Atabapo, nr. mouth of Río Atacavi, *Foldats 3697* (US, VEN); Dept. Atures, cercanía del Río Guayapo, al S de la Serranía de Sipapo, 4°13′N, 67°W, 120 m, *Huber 2479* (US, VEN); Serranía Parú, 4°7′N, 65°32′W, 1100 m, *Huber 4433* (VEN); Dept. Río Negro, Aracamuni Massif, 01°23′N, 65°48′W, ca. 750 m, *Huber & Medina 5897* (US, VEN); El Venado on left bank of Caño Pimichín, above Pimichín, 140 m, *Maguire & Wurdack 35560* (NY, VEN); Cerro Moriche, Río Ventuari, 800 m, *Maguire et al. 30890* (F, G, MO, NY, RB, US, VEN); around San Carlos de Río Negro, camino a Solano, *Stergios & Aymard 4015* (TFA).

It was initially thought that two subspecies, one restricted to the summits and one distributed in the lowlands, could be distinguished on the basis of glandularity; subsequent collections have shown that there is random variation of this character among the specimens.

18. Macairea philipsonii Renner, sp. nov. Type: Colombia. Meta: Sierra de la Macarena, 6 Feb 1950 (fl), *Philipson 2378* (holotype: COL!; isotype: BM!).

Planta suffruticosa tota densissime pilis glanduliferis hirta; calycis lobi angusti triangulares tubum aequales vel paulo longiores.

Subshrub; branchlets, petioles, and peduncles hirsute with glandular hairs. *Leaves* ovate or elliptic, 1.6–2.4 × 0.6–1.2 cm, acute, obtuse at base, 3-nerved, the pair of lateral primary veins arising at the blade base, above and below hirsute with glandular hairs, hairs on the lower surface slightly curled; petioles 0.3–0.5 mm long. *Inflorescence* a many-flowered terminal paniculate cyme. *Flowers* with pedicels 4–7 mm long; pedicels, bracts, hypanthium, and calyx lobes hirsute with glandular hairs; hypanthium 2–3 × 1.8–2.2 mm; calyx lobes narrowly triangular, 3–4 mm long; petals purple, 9–10 × 7–8 mm; filaments 4.8–5.2 or 3.8–4.2 mm long, glabrous; thecae ca. 2 mm long, the connective distinctly prolonged and expanded dorsi-basally; ovary apically with glandular hairs, 4-locular. *Fruit* capsular. *Seeds* n.v.

Distribution and ecology (Fig. 37A): *Macairea philipsonii* is only known from the Serranía de la Macarena, the westernmost outlier of the Roraima formation, 50 km from the Andes, where the habitat is reported as "open grassland on sandstone bluff" (*Philipson 2378*).

Macairea philipsonii has the dorsi-basal connective appendage and conspicuous glandular hair-covering characteristic of *Macairea*. Within the genus, it belongs to the large group of species which have long glandular hairs, but no sessile glands, and within that group it seems closest to *M. multinervia* and *M. lasiophylla* with which it shares the suffrutescent habit. The three species also share the lack of glands on the filaments and the weak separation of the vegetative and reproductive zones, with foliage leaves present in the lower part of the inflorescences. *Macairea philipsonii* differs from *M. multinervia* in the much smaller leaves and from *M. lasiophylla* in the larger flowers which are completely purple, a highly unusual trait in *Macairea.*

19. Macairea multinervia Bentham, J. Bot. (Hooker) **2:** 219. 1840; *Tetrameris trivalvis* Naudin, Ann. Sci. Nat. Bot. III, **14:** 121. 1850b; *Tibouchina multinervia* (Bentham) Baillon, Adansonia **12:** 75. 1876–1879. Type: "Guyana." Moist savannas of Roraima, 1842–1843 (fl), *Ri. Schomburgk 706/1073* (holotype: K!; isotypes: BM!, CAS!, G!, OXF!, P!, F photo neg. 36141!, W!; isotype at B destroyed, but represented by a photo, F neg. 16700!). Fig. 27K, L.

Subshrub or *shrub,* 0.2–1 m tall; branchlets, petioles, and peduncles hirsute with long, glandular hairs. *Leaves* ovate 3–4 × 2.5–3 cm, obtuse or acute, cordate at base, 7–9-nerved, the lateral primary veins arising at the blade base, both surfaces densely hirsute with glandular hairs, the areoles with shorter glandular hairs; petioles 0.3–0.8 cm long. *Inflorescence* a many-flowered, elongate, terminal paniculate cyme. *Flowers* with pedicels 3–6 mm long; pedicels, bracts, hypanthium, and calyx lobes hirsute with glandular hairs; hypanthium 3–4 × 2.0–2.5 mm; calyx lobes narrowly triangular, 2–4 mm long; petals basally white, turning red with age, apically magenta, 12–14 × 8–10 mm; filaments 8–10 or 7–8 mm long, glabrous; thecae 5–6 mm long, the connective prolonged and expanded dorsi-basally; ovary apically with glandular hairs, 3(–4)-locular. *Fruit* capsular. *Seeds* 0.4–0.6 mm long, subcochleate.

Distribution and ecology (Fig. 35B): *Macairea multinervia* has a disjunct distribution, occurring in the Gran Sabana in Venezuela and in savannas near Humaitá on the upper Rio Madeira in Brazil, 1200 km to the south; lowlands to 1300 m.

Representative specimens examined. VENEZUELA. BOLIVAR: Distr. Roscio, ca. 15 km SW de S. Ignacio de Yuruarí, 4°55′N, 61°14′W, 1200 m, *Huber & Alarcon 7585* (US); Ilú-tepuí, at Kavanayén, 1300 m, *Maguire 33670* (COL, F, G, NY); vic. of Urimán, 200 m, *Steyermark 75268* (F, NY, VEN); Río Urimán, just above Salto Acarima, 393 m, *Steyermark & Wurdack 36* (F, G, NY, RB, US, VEN); ca. 4 km E of Mission Sta. Teresita de Kavanayén, 1300 m, *Maguire & Wurdack 33995* (MO, NY, RB).

BRAZIL. AMAZONAS: Humaitá-Jacarecanga, 62 km S of km 150, *Teixeira et al. 1327* (INPA, US); *Cid et al. 5516* (MG, US).

As pointed out above under *Macairea parvifolia,* the types of that species and of *M. multinervia* were probably collected near the Venezuelan/Brazilian border rather than in Guyana. The leaves of *Macairea multinervia* differ from all other species in being thinner, basally cordate, and 7–9-nerved; however, the flowers of *M. multinervia* are those of a typical member of the genus.

20. Macairea lasiophylla (Bentham) Wurdack, Phytologia **13:** 65. 1966. Figs. 24F; 29E, F.

Chaetogastra ? lasiophylla Bentham, J. Bot. (Hooker) **2:** 291. 1840; *Pterolepis lasiophylla* (Bentham) Triana, Trans. Linn. Soc. London **28:** 40. 1871; *Tibouchina lasiophylla* (Bentham) Cogniaux *in* Martius, Fl. bras. **14(3):** 297. 1885; *Acisanthera lasiophylla* (Bentham) Gleason, Phytologia **3:** 243. 1950b. Type: Guyana. May 1842 (fl), *Ri. Schomburgk 568* (*=200.5*) (holotype: K!; isotypes: BR frag.!; B destroyed, but represented by photos, H. A. Gleason neg. 30-5! and F neg. 16759!).

Acisanthera erecta Gleason, Bull. Torrey Bot. Club **58:** 412. 1931a. Type: Venezuela. Terr. Fed. Amazonas: Upper Grand Savanna, nr. Monkey valley, Esmeralda, 4 Nov 1928 (fl), *Tate 332* (holotype: NY!; isotype: US!).

Siphantheropsis williamii Brade, Inst. Nac. Pesq. Amazônia Publ. Bot. **8:** 6. 1958. Type: Brazil. Roraima: Tributary of Rio Canã, 22 Dec 1954 (fl), *Rodrigues 142* (holotype: INPA!).

Herb or *subshrub,* 0.1–0.5 m tall; branchlets, petioles, and peduncles hirsute with glandular hairs. *Leaves* narrowly elliptic, 0.6–1.6 × 0.3–0.8 cm, acute, rounded at base, 5-nerved, the lateral primary veins arising at the blade base, above and below hirsute with glandular hairs to 1.5 mm long; petioles 0.3–0.8 cm long. *Inflorescence* a narrow, terminal paniculate cyme. *Flowers* with pedicels 2–3 mm long; pedicels, bracts, hypanthium, and calyx lobes hirsute with glandular hairs; hypanthium 2.5–2.8 × 2–3 mm; calyx lobes broadly triangular, 1.5–2.3 mm long; petals white with pink blotches or basally white, turning red with age, apically purple, 5–6–8 × 4–5 mm; filaments 2–3.7 or (1.7–)2.7(–3.2) mm long, glabrous; thecae 1.3–2.5 mm long, the connective minutely expanded dorsi-basally, ventrally with two ascending lobes; ovary with short glandular hairs, 3-locular; style subclavate. *Fruit* capsular. *Seeds* ca. 0.5 mm long, cochleate with dorsal ridges.

Distribution and ecology (Fig. 35D): *Macairea lasiophylla* is a widespread, sometimes weedy, species of sandy savannas in southwestern Venezuela and northern Brazil; from 100 to 1200 m elevation.

Representative specimens examined. VENEZUELA. AMAZONAS: Dept. Río Negro, Serranía Vinilla, ca. 20 km SW de Maracá, 2°26′N, 65°20′W, ca. 420 m, *Huber 6207* (US, VEN); Dept. Atabapo, S–SE of Sta. Bárbara del Orinoco, along rd. to S. Antonio del Orinoco, 3°45′N, 67°04′W, 100 m, *Huber & Tillett 2867* (US, VEN); Canaripo, Río Ventuari, 4°05′N, 66°50′W, 125 m, *Steyermark & Redmond 112793* (MO, US, VEN); Esmeralda, 100 m, *Tate 332*(NY, US). BOLIVAR: Nr. Canaima, 500 m, *Hertel et al. 15342* (M); Distr. Piar, al E del Churí-tepuí, valle del Río Karuay inferior, 5°18′N, 61°55′W, 900 m, *Huber 9783* (US); Distr. Roscio, Parupa, 5°43′N, 61°32′W, 1250 m, *Huber et al. 7200* (VEN); 2–4 km NW of Kamarata, 450 m, *Koyama & Agostini 7289* (F, NY); Serra do Sol, 1200 m, *Maguire & Maguire 40378* (US, VEN); between Kun and waterfall at Rué-merú (trib. of Kukenán), S of Mt. Roraima, 1065 m, *Steyermark 59127* (F, VEN); vic. of camp at km 150 in valley of savanna of Río Uarama, below Uarama-tepuí, NE of Luepa, 1200 m, *Steyermark & Nilsson 626* (US, VEN).

GUYANA. Southern Pakaraima Mts., Ando savanna, Sukabi river, affl. Ireng river, 800 m, *Maguire et al. 46190A* (NY).

BRAZIL. RORAIMA: Along Boa Vista BV 8 rd. (BR 174), km 200, 560 m, *Coradin & Cordeiro 891* (IAN, NY, US); Rio Quino, *Lützelburg 21511* (M).

The purple, ventrally bilobed connective appendages of *Macairea lasiophylla* are unique within the genus. However, in habit and in the glandular pubescence of the whole plant *M. lasiophylla* agrees well with other members of *Macairea,* especially with *M. multinervia* and *M. philipsonii.* The generic placement of this unusual species has long been a problem; the only two melastome genera which have ventral connective appendages similar to those of *M. lasiophylla, Acisanthera,* and *Ernestia,* differ in the ovary indument, the shape of the sepals, and the color of the appendage. I agree with Wurdack (1966) who considered that *M. lasiophylla* "represents a reduced evolutionary offshoot related to *M. multinervia.*"

21. Macairea stylosa Triana, Trans. Linn. Soc. London **28:** 1871. Type: Brazil. Amazonas: Casiquiare, Vasiva, and Pacimoni rivers, 1853–1854 (fr), *Spruce 3195* (holotype: K!, photo H. A. Gleason s.n.!; isotypes: BM!, BR!, G!, G-DC!, GOET!, LE!, RB!, W!).

Figs. 23D, 24C, 26D–F.

Shrub or *treelet,* 1–3(–5) m tall; branchlets glabrous. *Leaves* narrowly elliptic, 3–12 × 1–2.5 cm, apically rounded, cuneate or angustate at base, 5-pli-nerved, the inner pair of lateral primaries arising ca. 4 mm above the blade base, the outer pair close to the margin and inconspicuous, above glabrous and smooth, the midvein and lateral primary veins with few short glandular hairs below, the areoles with minute, few-branched, papillae-like hairs; petioles 0.1–0.4 cm long. *Inflorescence* 3–5-flowered paniculate cymes in the axils of the uppermost vegetative leaves; peduncle glabrous. *Flowers* with pedicels 10–17 mm long; hypanthium 4.7–5.2 × 2.5 mm; calyx lobes subulate, 3–5 mm long; petals white or pink, 11–13 × 9–11 mm; filaments 6–8 or 3.8–4.2 mm long, ventrally with short-stalked glands; thecae 4(–6) mm long, the connective prolonged and expanded dorsi-basally; ovary apically with glandular hairs, 4-locular. *Fruit* capsular. *Seeds* ca. 1 mm long, subcochleate to polyhedric.

Distribution and ecology (Fig. 37B): *Macairea stylosa* is endemic to the Orinoco-Río Negro drainage where it grows in wet savannas and is sometimes completely inundated during the high-water season.

Representative specimens examined. VENEZUELA. AMAZONAS: San Carlos de Río Negro, ca. 20 km S of confl. of Río Negro & Brazo Casiquiare, 1°56′N, 67°03′W, 119 m, *Clark & Maquirino 8234* (MO, NY); Sta. Cruz, margin of Río Atabapo, nr. mouth of Río Atacavi, *Foldats 3845* (US, VEN); Dept. Río Negro, margin of lower Río Siapa, *Huber & Medina 5769* (US, VEN); laguna of Caño Cotúa, W base of Cerro Yapacana, 3°68′N, 66°52′W, 100 m, *Huber & Kral 70698* (TFA); Caño Pimichín, below Pimichín, 125–135 m, *Maguire & Wurdack 35605* (B, G, NY, P); Río Pacimoni, 50 km above mouth, 100–140 m, *Maguire et al. 37600* (NY, US); NW base of Cerro Yapacana, 125 m, *Maguire et al. 41482* (B, COL, M, NY, P, S, US); Maroa, Río Guainía, 127 m, *L. Williams 14298* (F, G, NY, US, VEN).

BRAZIL. AMAZONAS: Rio Negro, São Felipe, *Fróes 28679* (RB); Rio Uneiuxi, 200–300 km above mouth, *Prance 15537* (COL, F, NY, S, US).

Macairea stylosa is illustrated in Cogniaux (1858, pl. 59, fig. 2). In almost all the specimens I have seen, the hypanthia were greatly enlarged due to infestation by the larvae of a gall-causing insect.

22. Macairea axilliflora Wurdack, Mem. New York Bot. Gard. **10:** 99. 1958b. Type: Venezuela. Terr. Fed. Amazonas: At edge of savanna on right bank of Río Pacimoni, 50 km above mouth, 100–140 m, 7 Feb 1954 (fr, old fl), *Maguire, Wurdack & Bunting 37601* (holotype: NY!, NY photo neg. 5645!; isotypes: S!, US!). Fig. 24A, B.

Shrub, 0.3–2 m tall; branchlets glabrous. *Leaves* narrowly obovate, 2.5–5 × 1–2 cm, acute, angustate at base, 3(–5)-nerved, the lateral primary veins arising at the blade base, the outermost pair close to the margin and inconspicuous, glabrous and smooth above, lower surface densely covered with minute sessile glands; petioles 0.1–0.5 cm long. *Inflorescence* consisting of groups of 3–5 subsessile flowers in the axils of the uppermost leaves. *Flowers* with hypanthium 2–3 × 1.8–2.2 mm, glabrous; calyx lobes triangular, ca. 1 mm long, with fine wooly hairs; petals pale pink or lilac, 3–4 × 1.8–2.2 mm; filaments 4.5–5.2 or 3.8–4.2 mm long, glabrous; thecae ca. 2 mm long, the connective minutely prolonged and expanded dorsi-basally; ovary apically with short glandular hairs, 2-locular. *Fruit* capsular. *Seeds* ca. 0.8 mm long, subcochleate to polyhedric.

Distribution and ecology (Fig. 35D): *Macairea axilliflora* is an endemic of the low-lying, seasonally flooded savannas of the Ríos Pacimoni and Atabapo; it has also been collected in secondary vegetation at the edges of clearings.

Representative specimens examined. COLOMBIA. GUAINIA: 1–2 km W of Río Guainía along caño, 0.5 km N of Boca de Casiquiare, 1°57′N, 67°08′W, 120 m, *Liesner & Clark 9121* (US, VEN).

VENEZUELA. AMAZONAS: Santa Cruz, margin of Río Atabapo, nr. mouth of Río Atacavi, *Foldats 3632* (US, VEN); Dept. Atabapo, 5 km S of Caserío Puruname, lower Río Puruname, 3°22′N, 66°30′W, 100 m, *Huber 5077* (NY, US, VEN); middle Río Puruname, 3°20′N, 66°32′W, 100 m, *Huber 5457* (VEN); Dept. Río Negro, lower Río Siapa, 01°57′N, 66°02′W, 125 m, *Huber & Medina 5757* (US, VEN); W bank of Río Casiquiare, 300 m below mouth of Río Pacimoni, *Maguire & Wurdack 35724* (G, NY, US); Bruno (Guaibana), Río Pacimoni, 1°40′N, 66°35′W, 100 m, *Steyermark & Bunting 102486* (F, MO, NY, US, VEN); Sta. Cruz, Río Atabapo, 1 km N of the confl. of the Río Temeni and the Río Atacavi, *Thomas & Rogers 2671* (NY).

Macairea axilliflora has numerous characters I consider derived within the genus: the lower leaf surface is completely covered with minute glands appearing like granules at low magnification; the inflorescences consist of subfasciculate clusters of small flowers in the axils of the uppermost foliage leaves; the ovary is 2-locular, and the filaments are glabrous. The wood of *M. axilliflora* differs in the number and diameter of the vessels from that of the other species investigated (Welle & Koek-Noorman, 1981, cf. *Habit, Wood, and Bark*).

Excluded Taxa

Macairea bracteolaris (Martius) Triana, Trans. Linn. Soc. Bot. **28:** 38. 1871.
=*Marcetia bracteolaris* (Martius) Cogniaux *in* Martius, Fl. bras. **14(3):** 441. 1885.

Acknowledgements

I am grateful to Dr. John J. Wurdack for a critical review of the *Bellucia* manuscript and to him and Dr. Harold Robinson for discussion, advice and helpful comments in the course of these studies. Drs. W. Rodrigues, M. Freitas da Silva, and M. L. Absy are thanked for permission to work at the Instituto de Pesquisas da Amazônia in Manaus (INPA, Brazil) for a total of about three years. Part of the fieldwork for the *Bellucia* study was done within the INPA-World Wildlife Fund "Minimal Critical Size of Ecosystems Project" and I thank the project's field directors for logistical support.

Thanks for fine technical assistance go to Janice L. Bittner, Susan Braden, and Heidi Wolf of the Palynological and Scanning Electron Microscope Laboratories of the National Museum of Natural History, Washington, D.C. The illustrations were provided by Cathy Pasquale and Alice Tangerini. Kenneth Wurdack grew the seedlings of *Macairea* species.

Thanks are extended to the curators of the following herbaria who put material at my disposal: AAU, B, BHMH, BM, BR, C, CAS, CM, COL, DS, F, G, GB, G-DC, GH, GOET, HB, HBG, HUA, IAN, INPA, JAUM, K, LBP, LE, LISU, LOJA, M, MG, MO, NA, NY, OXF, P, P-Lamarck, PORT, QCA, R, RB, S, SEL, SP, SPF, TFA, U, UB, UFAC, UFMT, US, VEN, W, and YU.

A Smithsonian postdoctoral fellowship funded the *Macairea* study in whole and the *Bellucia* and *Loreya* studies in part. The *Bellucia* work was supported in part by grants from the Deutsche Forschungsgemeinschaft.

Literature Cited

Atramentowicz, M. 1982. Influence du milieu sur l'activité locomotrice et la reproduction de *Caluromys philander* (L.). Rev. Ecol. (Terre Vie) **36:** 373–395.

Aublet, J. B. C. F. 1775. Histoire des plantes de la Guiane françoise. 4 vols. London, Paris.

Baas, P. 1981. A note on stoma types and crystals in the leaves of Melastomataceae. Blumea **27:** 475–479.

Baillon, H. E. 1876–1879. Nouvelles observations sur les Mélastomacées. Adansonia **12:** 70–101.

———. 1877. Histoire des plantes: Monographie des Mélastomacées. Vol. 7. L. Hachette et Cie., Paris, London, Leipzig.

Bentham, G. 1840. Enumeration of plants collected by Mr. Schomburgk in British Guiana. Melastomaceae. J. Bot. (Hooker) **2:** 286–315.

——— **& J. D. Hooker.** 1867. Genera plantarum **1:** 725–773.

Bonpland, A. J. A. 1820. Rhexiae. *In* A. Humboldt & A. J. A. Bonpland, Monographia Melastomacearum. Vol. 2. Paris.

Brade, A. C. 1958. Melastomatáceas novas da região amazônica. Inst. Nac. Pesq. Amazônia Publ. Bot. **8:** 1–18.

———. 1959a. Melastomataceae novae VI. Arq. Jard. Bot. Rio de Janeiro **16:** 1–16.

———. 1959b. Melastomatáceas novas do estado de Goiás. Arq. Jard. Bot. Rio de Janeiro **16:** 29–33, 4 pls.

Brewer Carías, C. 1978. La vegetación del mundo perdido. Caracas, Venezuela.

Brown, N. E. [and collaborators]. 1901. Report on two botanical collections made by Messrs. F. K. McConnell and J. J. Quelch at Mount Roraima in British Guiana. Trans. Linn. Soc. London **6(1):** 1–107.

Candolle, A. P. de. 1828a. Melastomaceae. *In* Prodromus systematis naturalis regni vegetabilis **3:** 99–202.

———. 1828b. Mémoires sur la famille des Mélastomacées. Paris.

Carreira, L. M. M. 1976. Indice de ataque de insectos em folhas e frutos de *Bellucia grossularioides* Triana (Melastomataceae). Ci. Cult. **28:** 685–687.

Cavalcante, P. B. 1979. Frutas comestíveis da Amazônia III. Museu Goeldi, Belém.

Charles-Dominique, P. 1986. Inter-relations between frugivorous vertebrates and pioneer plants: *Cecropia,* birds and bats in French Guyana. Pages 119–135 *in* A. Estrada & T. H. Fleming (eds.), Frugivores and seed dispersal. Dr. W. Junk, Dordrecht.

Chenery, E. M. 1948. Aluminium in the plant world. Kew Bull. **1948:** 173–183.

Cogniaux, A. 1885. Melastomaceae. *In* C. F. P. de Martius, Flora brasiliensis **14(3):** 1–510.

———. 1888. Melastomaceae. *In* C. F. P. de Martius, Flora brasiliensis **14(4):** 1–655.

———. 1891a. Melastomaceae. *In* Alph. & C. de Candolle (eds.), Monographie phanerogamarum **7:** 1–1256.

———. 1891b. Primitiae Florae Costaricensis 1. Bull. Soc. Roy. Bot. Belgique **30:** 264.

———. 1896. Eine neue Melastomaceen-Species aus der Gattung *Macairea.* Bot. Centralbl. **66:** 369.

———. 1908 [1909]. Melastomataceae peruvianae. *In* I. Urban (ed.), Plantae novae andinae imprimis Weberbauerianae IV. Bot. Jahrb. Syst. **42:** 131–148.

———. 1912. Melastomatáceas, Cucurbitáceas e Orchidáceas. *In* F. C. Hoehne (ed.), Melastomáceas. Comm. Linh. Telegr. Estr. Matto Grosso, Publ. 10, Annexo 5, Bot. **3:** 1–198.

Cremers, G. 1983. Architecture végétative et structure inflorescentielle de quelques Melastomaceae guyanaises. Univ. Strasbourg, France.

Croat, T. B. 1978. Flora of Barro Colorado Island. Stanford University Press, Stanford, California.

Crueger, H. 1847. Melastomaceae insulae Trinitatis. Linnaea **20:** 99–116.

Dandy, J. E. 1938. The Sloane Herbarium. London.

Desvaux, N. A. 1825. *Apatitia.* Page 42 *in* W. Hamilton, Prodr. pl. Ind. occid.

Détienne, P. & J. Jacquet. 1983. Atlas d'identification des bois de l'Amazonie et des régions voisines. Centre Technique Forestier Tropical, Nogent-sur-Marne, France.

Dodson, C. H. & A. H. Gentry. 1978. Flora of the Río Palenque Science Center. Selbyana **4:** 1–628.

Don, D. 1823. An illustration of the natural family of plants called Melastomaceae. Mem. Wernerian Soc. **4:** 276–329.

Duarte, A., L. Enrique, I. Rodríguez S., H. Delgado G. & A. Roa Torres. 1979. Principales plantas útiles de la Amazonía Colombiana. Editora Guadalupe, Bogotá.

Ducke, A. 1922. Plantes nouvelles ou peu connues de la région Amazonienne. Arq. Jard. Bot. Rio de Janeiro **3:** 3–281.

Erdtman, G. 1960. The acetolysis method. A revised description. Svensk Bot. Tidskr. **54:** 561–564.

———. 1971. Pollen morphology and plant taxonomy.—Angiosperms. Hafner, New York.

Forero Pinto, L. E. 1980. Etnobotánica de las comunidades indígenas Cuna y Waunana, Chocó (Colombia). Cespedesia **9:** 115–325.

Fournet, J. 1978. Flore illustrée des phanérogames de Guadeloupe et de Martinique. Inst. National de la Recherche Agronomique, Paris.

Friese, H. 1899. Monographie der Bienengattung *Euglossa* Latr. Természetrajzi Füzetek **22:** 117–172.

Gleason, H. A. 1931a. Recent collections of Melastomataceae from Peru and Amazonian Brazil. Bull. Torrey Bot. Club **58:** 215–262.

———. 1931b. Botanical results of the Tyler-Duida expedition. Bull. Torrey Bot. Club **58:** 277–506.

———. 1932. A synopsis of the Melastomataceae of British Guiana. Brittonia **1:** 127–184.

———. 1933. Melastomataceae. *In* H. A. Gleason & A. C. Smith, Plantae Krukovianae. Bull. Torrey Bot. Club **60:** 379–396.

———. 1934. The genus *Macairea* DC. in northern South America. Bull. Torrey Bot Club **61:** 35–40.

———. 1935. Melastomaceae. *In* A. Pulle (ed.), Flora of Surinam **3:** 178–281. Koninklijk Koloniaal Instituut te Amsterdam.

———. 1950a. Observations on tropical American melastomes. Phytologia **3:** 345–360.

———. 1950b. The genus *Tibouchina* in southern Venezuela. Phytologia **3:** 238–243.

———. 1953. Melastomataceae. *In* B. Maguire, R. S. Cowan & J. J. Wurdack (eds.), The botany of the Guayana highland. Mem. New York Bot. Gard. **8:** 133–144.

———. 1958. Melastomataceae. *In* R. E. Woodson et al., Flora of Panama 7(3). Ann. Missouri Bot. Gard. **45:** 203–304.

Gmelin, J. F. 1791. Systema naturae. Vol. 2. Leipzig.

Gottschall, M. 1900. Anatomisch systematische Untersuchung des Blattes der Melastomataceen aus der Tribus Miconieae. Mém. Herb. Boissier **19:** 1–176, 3 pls.

Grisebach, A. H. R. 1864. Flora of the British West Indian Islands. London.

Hallé, F., R. A. A. Oldeman & P. P. Tomlinson. 1978. Tropical trees and forests: An architectural analysis. Springer-Verlag, Berlin, New York.

Hamilton, W. 1825. Prodromus plantarum Indiae occidentalis. London.

Helten, W. M. van 1923. *Bellucia axinanthera* Triana. Alg. Landbouwweekbl. Ned.-Indië **7:** 1970.

Hoehne, F. C. 1922. Melastomáceas. Mem. Inst. Butantan, Sec. Bot. **1(5):** 1–198.

Holmgren, P., W. Keuken & E. K. Schofield. 1981. Index herbariorum. Part I. The herbaria of the world. 7th ed. Regnum Veg. 106. Utrecht.

Hooker, J. D. (ed.). 1867. Icones plantarum. Vol. 11.

Howard, R. A. 1983. The plates of Aublet's Histoire des plantes de la Guiane françoise. J. Arnold Arb. **64:** 255–292.

Huber, J. 1909 [1910]. Mattas e madeiras amazônicas. Bol. Mus. Paraense Hist. Nat. **6:** 91–216.

———. 1914. Plantae Duckeanae austro-guyanenses. Bull. Soc. Bot. Genève **6:** 179–213.

Jussieu, A. L. de. 1789. Genera plantarum. Paris.

Karsten, G. K. W. H. 1861. Flora Columbiae. Vol. 1. Berlin.

———. 1859. *Axinanthera.* Page 157 *in* Plantae Karstenianae ex variis familiis. Linnaea **30:** 143–165.

Koek-Noorman, J., P. Hogeweg, W. H. M. van Maanen & B. J. H. ter Welle. 1979. Wood anatomy of the Blakeeae (Melastomataceae). Acta Bot. Neerl. **28:** 21–43.

Lemée, A. M. V. 1955. Flore de la Guyane francaise. Vol. 1., Paul Lechévalier, Paris.

———. 1956. Flore de la Guyane française. IV. Végétaux utiles de la Guyane francaise., Paul Lechévalier, Paris.

Lieberman, D., M. Lieberman, R. Peralta & G. S. Hartshorn. 1985. Growth rates and age-size relationships of lowland tropical wet forest in Costa Rica. J. Trop. Ecol. **1:** 97–109.

Linnaeus, C. 1753. Species plantarum. Stockholm.

———. 1762–1763. Genera plantarum. Ed. 2. Stockholm.

Linné, C. von, filius. 1781. Supplementum plantarum. Braunschweig.

Macbride, J. F. 1941. Melastomaceae. *In* Flora of Peru. Field Mus. Publ. Bot. **13:** 249–523.

Maguire, B. 1970. On the flora of the Guayana highland. Biotropica **2:** 85–100.

Mayr, E. & W. H Phelps, Jr. 1967. The origin of the bird fauna of the South Venezuelan highlands. Bull. Amer. Mus. Nat. Hist. **136:** 269–328.

McVaugh, R. 1982. The lost paintings of the Sessé & Moçiño expedition: A newly available resource. Taxon **31:** 691.

Mee, M. 1968. Flowers of the Brazilian forests, collected and painted by Margaret Mee. Tryon Gallery, London.

Metcalfe, C. R. & L. Chalk. 1979. Anatomy of the dicotyledons. Ed. 2, Vol. 1. Clarendon Press, Oxford.

Miquel, F. A. G. 1844. Symbolae ad floram surinamensem. Linnaea **18:** 562–624.

Mutis, J. C. B. (1983). Flora de la Real Expedición botánica del Nuevo Reino de Granada (1760–1817). Tomo 31 (Melastomatáceas segunda parte). Ediciones cultura hispánica, Madrid.

Naudin, C. V. 1850a. Melastomacearum monographicae descriptiones. Ann. Sci. Nat. Bot. III, **13:** 273–303.

———. 1850b. Melastomacearum monographicae descriptiones. Ann. Sci. Nat. Bot. III, **14:** 118–165.

———. 1850c. Melastomacearum monographicae descriptiones. Ann. Sci. Nat. Bot. III, **16:** 83–246.

———. 1852. Melastomacearum monographicae descriptiones. Ann. Sci. Nat. Bot. III, **18:** 86–154.

Necker, N. J. de. 1790. Elementa botanica. 3 vols. Neuwied.

Nettancourt, D. de. 1977. Incompatibility in angiosperms. Springer, Berlin.

Olivier, G. H. 1789. Encyclopedie methodique de histoire naturelle . . . Insectes. Liège 4. Paris. 331 pp.

Patel, V. C., J. J. Skvarla & P. H. Raven. 1985. Pollen characters in relation to the delimitation of Myrtales. Ann. Missouri Bot. Gard. **71:** 858–969.

Pflaum, F. 1897. Anatomisch-systematische Untersuchung des Blattes der Melastomataceen aus den Triben Microlicieen und Tibouchineen. Doctoral Dissertation. University of Munich. Pp. 1–91.

Pilger, R. K. F. 1902 (1901). Beitrag zur Flora von Matto Grosso. Bot. Jahrb. Syst. **30:** 127–238.

———. 1905. Melastomataceae. Verh. Bot. Vereins Prov. Brandenburg **47:** 165–186.

Pittier, H. F. 1923. Exploraciones botánicas y otras en la Cuenca de Maracaibo. Tipografía mercantil, Caracas.

———. 1926. Manual de las plantas usuales de Venezuela. Caracas.

Plukenet, L. 1692. Phytographia. Pars 3. London.

Plumier, C. 1755–1760. Plantarum americanarum. Leiden.

Rafinesque, C. S. 1838. Sylva telluriana. Philadelphia.

Rehder, A., C. A. Weatherby, R. Mansfeld & M. L. Green. 1935. Conservation of later homonyms. Bull. Misc. Inform. **6:** 341–544.

Renner, S. S. 1984a. Phänologie, Blütenbiologie und Rekombinationssysteme einiger zentralamazonischer Melastomataceen. Doctoral Dissertation. University of Hamburg, West Germany [printed copy].

———. 1984b. Pollination and breeding systems in some Central Amazonian Melastomataceae. Proc. 5th Intern. Symp. Pollination, INRA publ.: 175–280.

———. 1986/1987. Reproductive biology of *Bellucia* (Melastomataceae). Acta Amazonica **16/17:** 197–208.

———. In press. A survey of reproductive biology in neotropical Melastomataceae and Memecylaceae. Ann. Missouri Bot. Gard.

Romero-Castañeda, R. 1961. Frutas silvestres de Colombia. San Juan Eudes, Bogotá.

Roosmalen, M. G. M. van. 1985. Fruits of the Guianan flora. Inst. Syst. Botany, Utrecht Univ., Netherlands.

Sagot, P. A. 1883. Catalogue des plantes . . . de la Guyane francaise. Ann. Sci. Nat. Bot. VI, **15:** 303–336.

Schomburgk, Ri. 1847–1848. Reisen in Britisch-Guiana in den Jahren 1840–1844. 3 vols. Leipzig.

Seeman, B. C. 1854. The botany of the voyage of H.M.S. Herald under the command of Captain Henry Kellett . . . during the years 1845–1851. London.

Solereder, H. 1908. Systematic anatomy of the dicotyledons. 2 vols. Transl. by L. A. Boodle & F. E. Fritsch. Clarendon Press, Oxford.

Solt, M. L. & J. J. Wurdack. 1980. Chromosome numbers in the Melastomataceae. Phytologia **47:** 199–220.

Spruce, R. 1908. Notes of a botanist on the Amazon and the Andes during the years 1849–1864. Vol. 2. London.

Standley, P. C. 1929. Studies of American plants—I. Publ. Field Mus. Nat. Hist., Bot. Ser. **4:** 197–299.

———. 1938. Melastomataceae. Flora of Costa Rica. Publ. Field Mus. Nat. Hist., Bot. Ser. **18:** 783–845.

———. 1961. Trees and shrubs of Mexico. Smithsonian Contr. Bot. 23.

——— **& L. O. Williams.** 1963. Melastomataceae. *In* Flora of Guatemala. Fieldiana, Bot. **24:** 407–525.

Steenis, C. G. G. J. van. 1975. Miscellaneous botanical notes 23. Blumea **22:** 169.

Steyermark, J. A. 1979. Flora of the Guayana Highland: Endemicity of the generic flora of the summit of the Venezuela tepuis. Taxon **28:** 45–54.

St. John, T. 1980. Uma lista de espécies de plantas tropicais brasileiras naturalmente infectadas com micorriza vesicular-arbuscular. Acta Amazonica **10:** 229–234.

Stone, B. C. 1972. Discovery in West Malaysia of *Bellucia,* a cultivated fruit tree from South America. Planter, Kuala Lumpur **48:** 276–278, 1 fig.

Tobe, H. & P. H. Raven. 1983. An embryological analysis of Myrtales: Its definition and characteristics. Ann. Missouri Bot. Gard. **70:** 71–94.

Triana, J. 1871. Les Melastomacées. Trans. Linn. Soc. London **28:** 1–188.

Urban, I. 1920. Plumiers Leben und Schriften nebst einem Schlüssel zu seinen Blütenpflanzen. Rep. Spec. Nov. Regni Veg. Beih. **5:** 1–196.

Uribe Uribe, L. 1972. Catálogo ilustrado de las plantas de Cundinamarca. Vol 5. Passifloraceae, Begoniaceae, Melastomataceae. Inst. Ciencias Naturales, Bogotá.

Vliet, G. J. C. M. van, J. Koek-Noorman & B. J. H. ter Welle. 1981. Wood anatomy, classification and phylogeny of the Melastomataceae. Blumea **27:** 463–473.

Vogel, S. 1963. Duftdrüsen im Dienste der Bestäubung. Über Bau und Funktion der Osmophoren. Akad. Wiss. Lit. Mainz, Math.-Nat. Klasse **10:** 601–763.

———. 1966. Parfümsammelnde Bienen als Bestäuber von Orchidaceen und *Gloxinia.* Österr. Bot. Zeitschr. **113:** 302–361.

Welle, B. J. H. ter & A. M. W. Mennega. 1977. On the presence of large styloids in the secondary xylem of the genus *Henriettea* (Melastomataceae). IAWA Bull. **2:** 31–35.

——— **& J. Koek-Noorman.** 1978. On fibres, parenchyma and intermediate forms in the genus *Miconia* (Melastomataceae). Acta Bot. Neerl. **27:** 1–9.

——— **&** ———. 1981. Wood anatomy of the neotropical Melastomataceae. Blumea **27:** 335–394.

Whiffin, T. & A. S. Tomb. 1972. The systematic significance of seed morphology in the neotropical capsular-fruited Melastomataceae. Amer. J. Bot. **59:** 411–422.

Wilson, C. L. 1950. Vasculation of the stamens in the Melastomataceae, with some phyletic implications. Amer. J. Bot. **37:** 431–444.

Wurdack, J. J. 1958a. Melastomataceae. *In* R. E. Schultes, Plantae austro-americanae X. Bot. Mus. Leafl. **18:** 160–166.

———. 1958b. Melastomataceae. *In* B. Maguire, J. J. Wurdack, and collaborators, The botany of the Guayana Highland—Part 3. Mem. New York Bot. Gard. **10:** 95–117.

———. 1962. Melastomataceae of Santa Catarina. Sellowia **14:** 109–217.

———. 1964. Melastomataceae. *In* B. Maguire, J. J. Wurdack, and collaborators, The botany of the Guayana Highland—Part 5. Mem. New York Bot. Gard. **10:** 135–186.

———. 1965. Certamen Melastomataceis IX. Phytologia **11:** 377–400.

———. 1966. Certamen Melastomataceis X. Phytologia **13:** 65–80.

———. 1969. Certamen Melastomataceis XIII. Phytologia **18:** 147–163.

———. 1970. Certamen Melastomataceis XV. Phytologia **20:** 369–389.
———. 1973. Melastomataceae. *In* T. Lasser (ed.), Flora de Venezuela **8:** 1–819.
———. 1980a. Melastomataceae. *In* G. Harling & B. Sparre (eds.), Flora of Ecuador **13:** 1–406.
———. 1980b. Certamen Melastomataceis 31. Phytologia **45:** 321–335.
———. 1982. Certamen Melastomataceis 34. Phytologia **50:** 297–308.
———. 1986. Atlas of hairs for neotropical Melastomataceae. Smithsonian Contr. Bot. **63:** 1–80.

NUMERICAL LIST OF TAXA: *BELLUCIA*

1. *B. aequiloba* Pilger
2. *B. pentamera* Naudin
3. *B. dichotoma* Cogniaux
4. *B. grossularioides* (L.) Triana
5. *B. acutata* Pilger
6. *B. beckii* Renner
7. *B. huberi* (Wurdack) Renner

LIST OF EXSICCATAE: *BELLUCIA*

Numbers in parentheses following each collection refer to the species numbered above.

Ackerly, D. D., 160 (3).
Acosta M., A., 47 (1).
Adaír, s.n. (3).
Aguírre-Galviz, L. E., 934 (4).
Albuquerque, B. W. de et al., 638 (4); 1270 (1); 1277 (1).
Allen, P. H., 3669 (2).
Alverson, B. & S. White, 204 (2).
Amarál, I. L. et al., 1147 (4).
Anderson, A., x-1811 (4).
Anderson, W. R., 10511 (4); 11858 (5); 11919 (4); 11968 (2); 12307 (2).
André, E. F., 1214 (4).
Archer, W. A., 2539 (4); 2799 (4).
Arias, L. A., 015 (2).
Aristeguieta, L., 1586 (2); 4972 (4).
Aristeguieta, L. et al., 7194 (2).
Asplund, E., 14027 (2); 19446 (2).
Asunción Leveau, J., 206 (2).
Aublet, J. B. C. F., s.n. (4).
Aubréville, A., 117 (4); 292 (4); 298 (4).
Austin, D. F. et al., 7022 (4).
Ayala, F. et al., 2812 (2); 3253 (4).
Aymard, G. et al., 3851 (4).
BAFOG, 53N-1082 (4).
Bahia, T. R., 5 (4); 15 (4); 83 (5).
Baker, M. A., 6154 (2).
Baker, M. A. et al., 6340 (2).
Balslev, H., 2305 (2).
Balslev, H. & R. Alarcon, 3030 (2).
Bamps, P., 5460 (5).
Bang, M., 1645 (1).
Barclay, A. S., P. Juajibioy & J. Gama, 3192 (2).
Barkley, F. A., 38C592 (2).
Barrier, S., 597 (2); 1811 (2).
Bartlett, H. H. & T. Lasser, 16351 (2).
Beck, S., 4904 (1).
Beck, S. & R. Haase, 9999 (4); 10000 (6).
Belshaw, C. M., 3384 (2).
Béña, R., s.n. (4).
Benítez de Rojas, C. E., 1221 (2).
Benoist, R., 1227 (4).
Berlin, B., 314 (2); 842 (2).
Bernal M., H. Y. & D. Tellez, 491 (2).
Bernardi, L., 2632 (4); 6581 (4); 6704 (4).
Bernoulli, C. G. & Cario, 2865 (4).
Berry, P., 715 (4).
Berti, L. M. & J. Bautista B., 33-6-78 (2).
Billiet, F. P. & B. Jadin, 1494 (4); 1993 (4).
Black, G. A., 46-7 (2); 47-1112 (3); 48-3317 (2); 54-16235 (4); 55-18487 (3).
Black, G. A. & E. Cordeiro, 52-14527 (5).
Black, G. A. & R. Foster, 48-3420 (2).
Black, G. A., R. L. Fróes & P. Ledoux, 50-9834 (4).
Blanco C., C. H., 601 (4); 1151 (4).
Blum, K. E., J. Dwyer & H. Loftin, 2532 (2).
Bohlin, J.-E. & M. Bohlin, 288 (2).
Bonpland, A. J. A., s.n. (4).
Boom, B. M., 2536 (2); 4285 (4); 4333 (4); 4404 (4); 4490 (1); 4534 (4).
Brand, J. & R. Marín, s.n. (2).
Brandbyge, J. & E. Asanza C., 31948 (2); 32088 (2).
Brandbyge, J., E. Asanza C. & R. Akkermans, 30212 (2); 30680 (2).
Brantjes, N. B. M., 700401 (4).
Breteler, F. J., 4773 (4).
Bruijn, J. de, 1200 (2); 1334 (2); 1453 (4).
Buchtien, O., 1033 (1); 1687 (1); 1729 (1); 1738 (2).
Bunting, G. S., 2814 (2); 5019 (2); 9415 (4); 10782 (4); 10849 (2).
Bunting, G. S. & L. Licht, 1041 (2).
Burger, W. & G. Metta U., 4825 (2).
Buschbacher, R. J., 82 (4).
B.W., 1860 (4); 1985 (2); 2508 (2); 2655 (2); 2725 (2); 3297 (4); 3824 (4); 4917 (4); 5005 (2); 5571 (2); 5968 (4).
Cabrera, I., 2505 (4); 3318 (4); 3715 (4).
Calderon, C., 2514 (4).
Calléjas, P., R., 209 (2).
Calzada, J. I., 7358 (2).

Campbell, W. H., 6/71 (4).
Cardona, F., 98 (2); 2137 (4).
Carleton, M. A., 546 (2); 579 (2).
Carreira, L., 95 (3); 144 (4).
Carreira, L., C. Moura & R. P. Lima, 857 (4).
Carreira, L. et al., 322 (4); 467 (5).
Castillo, A., 1468 (2).
Castillo C., G., 401 (2).
Cavalcante, P., 3127 (4); 3311 (3).
Cavalcante, P. & M. Silva, 1767 (3).
Chagas, J., 70 (4); 2208 (4); 3020 (4).
Chrostowski, M. S., 70-391 (2).
Chrysler, M. A. & W. E. Roever, 5232 (2).
Cid, C. A. & B. W. Nelson, 2634 (2); 2743 (2).
Cid, C. A. & J. Ramos, 1085 (4).
Cid, C. A. et al., 1831 (4); 2483 (4); 3987 (5); 4022 (5); 5196 (1); 6441 (4).
Clark, H. L., 6431 (4).
Clewell, H., 4510 (4).
Clewell, H. & G. Cruz, 4123 (2).
Clewell, H. & E. Tyson, 3226 (2).
Coêlho, D., O. Monteiro & J. Guedes, INPA 92478 (4).
Cogollo y Cruz, A. & C. Estrada, 275 (4).
Conceição, P. N., 2 (3).
Contreras, E., 5999 (4); 7591 (2); 8010 (4).
Cordeiro, M. R., 648 (4); 667 (5); 970 (4); 1071 (4).
Costich, D., 1052 (3).
Cowan, C., 2591 (4); 3121 (2); 3170 (4).
Cowan, R. S. & J. J. Wurdack, 32018 (4).
Cremers, G., 5252 (4); 8232 (4).
Croat, T., 7534 (2); 7644 (2); 12485 (2); 14979 (2); 16520 (2); 18179 (2); 21113 (2); 54127 (4); 58160 (2).
Croizat, L., 131 (4); 953 (4).
Cuadros V., H., 1531 (4).
Cuatrecasas, J., 3522 (2); 4646 (2); 6848 (4); 7323 (4); 7873 (4); 8843 (2); 8838 (4); 10553 (4); 13130 (2); 17119 (2); 21136 (2).
Cuatrecasas, J. & M. Llano, 24079 (2).
Cuatrecasas, J., T. Soderstrom & L. E. Mello Filho, 27242 (2).
Dahlgren, B. E. & E. Sella, 536 (4).
Daly, D. C. et al., 1624 (2); 3898 (4).
Damião, C., 2729 (2).
Daniel, Bro., 5537 (2).
Davidse, G., 5674A (2).
Davidse, G. & A. C. González, 13974 (4); 21648 (2).
Davidse, G. & J. S. Miller, 26535 (4).
Davidse, G. et al., 18047 (4); 18331 (2).
Davidson, C., 3452 (2).
Davidson, C. & G. Martinelli, 9972 (3); 10075 (4); 10646 (4).
Davis, D. M., 29 (4).
Davis, E. W., 115 (4).
Dawe, M. T., 239 (4).
Deam, C. C., 6050 (2).
Delascio, F., 4394 (2).
Delascio, F. & R. Lopez, 8763 (4).
Delascio, F., G. Christenson & C. Broome, 9457 (4).
De la Cruz, J. S., 2672 (4); 3254 (4).
Denslow, J., 2479 (2).
Díaz, 954 (suspected hybrid 2 × 4).
Díaz, C., H. Osores & N. Jaramillo, 696 (2).
Díaz, C. & J. Aronson, 730A (2).
Díaz P., S., 3413 (2).
Dodson, C. H. & J. Torres, 2926 (2).
Donselaar, J. van, 1398 (4).
Dorantes, J. et al., 2260 (4).
Dressler, R. L., 3499 (2).
Duarte, A. P., 6966 (4).
Duchassaing de Fontbressin, E. P., s.n. (4).
Ducke, A., 8 (3); 92 (3); 1353 (4); 2952 (3); 7155 (5); 7403 (4); 7574 (2); 11143 (4); 12041 (4); 12397 (5).
Duke, J. A., 5754 (4); 6553 (4); 8404 (2); 8926 (4); 11035 (2); 14734 (4).
Duke, J. A. & H. W. Mussell, 6553 (4).
Dunsterville, G. C. K., s.n., Oct 1969 (2).
Duss, Père, 1167 (4); 3205 (4).
Dwyer, J., 1789 (2); 5134 (2); 5195 (2).
Dwyer, J., T. Elias & R. Maxwell, 417 (4); 11354 (4).
Dwyer, J. et al., 5423 (2).
Echeverry E., P., 3427 (2).
Egler, W. A., 617 (2); 1403 (4); 47188 (4).
Elias, J., 332 (4).
Englesing, F. C., 109 (2).
Escobar, L. de et al., 2624 (2).
Espina, L., 689 (2); 1412 (2).
Espinal T., S. & E. Montenegro, 185 (2), 799 (2); 1501 (2); 1655 (2).
Estrela, M. R. & W. Bellusci, 88-5476 (5).
Ewan, J. A., 16753 (2).
Fagerlind, F. & G. Wibom, 2326 (2).
Fanshawe, D. B., 735 = FD 3471 (4).
Fassett, N. C., 25466 (4).
Fernández C., J. & Susanna, 8201 (suspected hybrid 1 × 2); 8205 (4); 8612 (4).
Fernández-Pérez, A., 20038 (2).
Ferrari, G., 1746 (2).
Ferreira, A. R., s.n., P sheet (4).
Ferreyra, R., 17444 (2).
Fonnegra, R., J. Santa & C. H. Torres, 632 (2).
Fonseca, S., 363 (2).
Forero, E., 670 (2).
Forero, E. & R. Jaramillo M., 9460 (2).
Forero, E. et al., 2423 (2); 4699 (2); 6379 (2); 8915 (2); 9460 (2).
Forest Dept. of British Guiana, FD 3471 (4); FD 5945 (4); FD 5979 (4).
Foster, R. B., 8517 (2).
Foster, R. & B. d'Achille, 10203 (2).
Franco, P. et al., 1091 (2).
Freitas, C. A. A. et al., 03 (4).
Fróes, R., 1722 (4); 23051 (5); 25150 (5); 31750 (3); 33181 (4); 34233 (2).
Fuchs, H. P. & L. Zanella, 22110 (2).
García-Barriga, H., 1157 (2); 3522 (4); 11215 (2); 12400 (2); 13142 (2); 16241 (4).
García-Barriga, H. & G. Lozano-C., 18275 (2).
Gasche, J. & J. Desplats, 99 (4); 1187 (4).
Gay, C., s.n., cat. no. 1269 (2).
Gehriger, W., 281 (2).
Gély, A. C., 121 (4).
Gentle, P. H., 1940 (4); 5587 (2); 6056 (4); 7745 (4).
Gentry, A., 6287 (2); 12713 (2); 20733 (2).
Gentry, A. & P. Berry, 14656 (4).

Gentry, A., G. Davidse & F. Llanos, 9104 (2).
Gentry, A., C. Diaz & N. Jaramillo, 21873 (2); 51255 (2).
Gentry, A. & M. Fallen, 17532 (2).
Gentry, A. & L. Forero, 15366 (4); 10857 (4).
Gentry, A. & A. Juncosa, 40556 (2).
Gentry, A. & H. León, 20217 (2).
Gentry, A. & M. Monsalve, 48182 (2).
Gentry, A., M. Monsalve & D. Wolfe, 47980 (2).
Gentry, A. & E. Rentería, 20076 (4).
Gentry, A. & J. Revilla, 16165 (2); 20548 (2).
Gentry, A., J. Solomon & E. Zardini, 44318 (2).
Gentry, A. & B. Stein, 47165 (4).
Gentry, A. et al., 18585 (suspected hybrid 2 × 4).
Gieteling, C. J., 82 (4).
Ginzberger, A., 613 (3).
Giovanni, F. di, s.n. (4).
Gleason, H. A., 323 (4).
Goeldi, A., 3868 (3).
Gomes, M., S. Mirando & C. D. Mota, 644 (4).
Gordon, B. L., 14 (2); 30C (2).
Gottsberger, G. & W. Morawetz, 15-5575 (5).
Graham, E. H., 410 (4).
Granville, J.-J., 2895 (4); 5122 (4); 6868 (4).
Grewal, M. S. & R. Persaud, 47 (2).
Guanchez, F., 168 (4); 243 (4).
Gutiérrez V., G. & R. E. Schultes, 712 (4).
Gutiérrez, O. & J. Sandoval, 09 (2).
Hallé, F., 1075 (4).
Hammel, B., 3402 (2).
Harling, G., 4303 (2).
Harling, G. & L. Andersson, 13767 (2); 13877 (2); 17407 (2).
Harmon, W. E. & J. A. Fuentes, 2048 (2).
Hay, R., s.n. (2).
Hayes, S., s.n., 18 Aug 1862 (4); 394 (4).
Hayward, R. B., 278 (4).
Hazlett, D., 3313 (4).
Hekking, W. H. A., 1042 (4).
Henderson, A. et al., 220 (5).
Hermann, F. J., 11288 (2).
Hernández G., H., 176 (2).
Hernández O., J. J., S. Hoyos M. & J. Restrepo, 128 (2); 577 (2).
Hitchcock, A. S., 17169 (4).
Hoehne, F. C., 5239 (4); 5240 (4).
Holdridge, L. R., 440 (2); 6606 (4).
Holm-Nielsen, L. et al., 20164 (2); 22110 (2); 22506 (2).
Holst, B. & R. Liesner, 3230 (7).
Holt, E. G. & W. Geriger, 281 (4).
Holt, E. G. & E. R. Blake, 634 (4).
Hostmann, F. W., 192 (4); 765a (4); 784b (4); 2742 (4).
Huashikat, V., 437 (2); 766 (2).
Huber, J., 123 (3); 1053 (4).
Huber, O. & E. Medina, 5652 (4).
Huber, O. & S. Tillett, 6398 (4).
Humbert, H. & A. Fernandez, 27209 (4).
Idrobo, J. M. & H. Weber, 1442 (2).
Ijjász-Madriz, 263 (2).
Irvine, D., 120 (2); 191 (2); 417 (2).
Irwin, H. S., 48711 (4); 48756 (3).
Irwin, H. S., H. Maxwell & D. C. Wasshausen, 21497 (2).
Irwin, H. S., J. M. Pires & L. T. T. Westra, 48449 (4).
Irwin, H. S. et al., 55573 (4).
Janssen, A., 251 (4); 272 (5).
Jaramillo M., R., 36 (4).
Jaramillo M., R. et al., 403 (4).
Játiva, C. & C. Epling, 2247 (2).
Jenman, G. S., 665 (4); 1925 (4); 4297 (4); 4549 (4); 4735 (4).
Jímenez, A., 867 (4); 3810 (2).
Jones, G. C., G. R. Proctor & L. Facey, 3018 (2).
Jonker-Verhoef, A. M. E. & F. P. Jonker, 246 (4).
Juncosa, A., 1330 (2).
Kalbreyer, W., 836 (2).
Kappler, A., 1461 (4); 2696 (4).
Karsten, H., s.n. (4); s.n. (2).
Kayap, R., 730 (2).
Keel, S., 293 (4); 353 (5).
Kegel, H. A. H., 337 (4).
Kellerman, W. A., 6108 (2).
Killip, E. P., s.n. (4).
Killip, E. P. & A. C. Smith, 24626 (2); 26898 (2); 27987 (2).
Killip, E. P., A. C. Smith & W. J. Dennis, 29215 (2).
Kirkbride, J. H. & E. Lleras, 2774 (4).
Klaas, B., 03 (5).
Knight, D. H., 691 (2).
Knob, A. et al., 465 (3).
Kramer, K. U. & W. H. A. Hekking, 3102 (4).
Krukoff, B. A., 1519 (4); 1572 (2); 1722 (4); 5580 (2); 7093 (5); 7096 (5); 8057 (2); 8567 (4).
Kuhlmann, J. G., 104 (4); 281 (4); 700 (2); 1083 (4); 1305 (2); 2474 (4).
Kuylen G., H., 146 (2).
Kuyper, J., 71 (4).
Kvist, L. P. & E. Asanza C., 40801 (2).
Labroy, M., W14 (3); s.n. (4).
LBB, 296 (4).
Lall, H., 354 (4).
La Rotta, C., 53 (2).
Ledoux, P., 2090 (2).
Lehmann, F. C., B. T. 559 (2); 7274 (2).
Lems, K., 650121 = 5171 (2).
Le Prieur, M., 295 (4).
Liesner, R. L., 3369 (4); 6547 (4); 4386 (4); 8799 (4); 19686 (4).
Liesner, R. L. & A. González, 9370 (2); 9702A (2); 10464 (2).
Liesner, R. L. & B. Holst, 18923 (4); 21343 (7).
Lima, D. A., 53-1636 (3).
Lindeman, J. C., 685 (4); 5911 (4); 6097 (4); 6191 (4); 6282 (4); 6462 (4).
Linden, J. J., s.n. (4); s.n. (2); 249 (2).
Lindsay, W. R., 232 (4).
Lisbôa, P., 207 (3).
Little, E. L., 6270 (2); 25141 (2).
Little, E. L. & R. R. Little, 8254 (4); 8265 (2); 9499 (4).
Little, E. L. et al., 713 (2).
Lizot, J., 75/12 (2); 161 (2).
Loaiza, C. & J. Hernandez, 342 (2).
Loaiza, C. A. et al., 174 (2).

Lobo, M. G. A., R. Vilhena & B. G. S. Ribeiro, 145 (3).
Lonzano C., G., 2298 (2).
Lonzano C., G. & R. Castillo, 410 (2).
López M., A., A. Sagástegui A. & F. Ayala, 8690 (2).
López P., S. & F. Bolívar, 718 (2).
López P., S., A. Caralot & C. Rosquete, 4591 (2).
Loureiro, A. et al., INPA no. 4790 (4); INPA no. 48480 (4).
Lowe, J., 4180 (4); 4249 (3).
Lowrie, S. R., B. Lowy & V. M. de Souza, 257 (2).
Lowrie, S. R. et al., 58 (4); 80 (5).
Lugo S., H., 1492 (2); 1530 (2); 1585 (2).
Lumer, C., 1363 (2).
Luteyn, J. L., 4893 (2).
Macbryde, B., 157 (2); 1493 (2); 1502 (2).
Madison, M. T. et al., 4801 (2); 5696 (4).
Madrigal, B. E. & G. Zuluaga, 176 (2).
Magnago, F. H. & G. Mota, 39 (3).
Maguire, B., 23785 (4); 23968 (4); 32635 (4).
Maguire, B. & C. K. Maguire, 62256 (4).
Maguire, B., R. S. Cowan & J. J. Wurdack, 29456 (4).
Maguire, B. et al., 54129 (4); 56703 (5).
Maia, L. A. et al., 657 (4).
Marcano-Berti, L. & I. Peña, 118-980 (4).
Marcano-Berti, L. & J. Bautista B., 33-6-78 (2).
Marinho, L. R., 63 (4).
Mathews, A., s.n. (2).
Martinelli, G., 6785 (4); 7290 (4).
Martius, C. F. P. von, s.n. (4); s.n. (5); 2770 (4).
Mattos, J., 10230 (4).
Matuda, E., 3712, mixed collection (2 and 4).
Mazuera, H., 25 (2).
McDaniel, S., 10910 (2); 16084 (2).
Mélinon, E., s.n. (4); 64 (4); 80 (4); 89 (4); 191 (4); 193 (4); 215 (4); 272 (4); 352 (4).
Mello, F. C., 2848 (4).
Mello, F. C., & L. Coêlho, 4166 (4).
Mennega, A. M. W., 11 (4).
Miranda, S., 07 (1).
Molina, J. A., 1949 (2); 14723 (2); 15140 (2).
Molina, J. A. & F. A. Barkley, 19Ch126 (2).
Molina, J. A. et al., 18004 (2); 18093 (2); 18232 (2).
Monteiro, O. P., INPA no. 50865 (5); INPA no. 50888 (4).
Monteiro, O. P. & C. Damião, 210 (4).
Monteiro, O. P. & J. Ramos, 880 (5); 881 (4).
Mora, L. E., 2262 (2).
Mori, S. A., 822 (3).
Mori, S. A. & C. Gracie, 18411 (4).
Morillo, G., 6723 (4); 7480 (4).
Morillo, G. & M. Hasegawa, 5062 (4).
Morillo, G. & M. Ishikawa, 3564 (4).
Morillo, G., E. Rutkis & G. Panapera, 9188 (2).
Mota, C. D., INPA no. 60439 (4).
Mota, C. D. & L. Coêlho, 103 (5); 116 (5).
Mutis, J. C., 3488 (2).
Nascimento, J. R., 308 (3).
Nascimento, J. R. & C. Damião, INPA no. 66185 (3).
Nascimento, J. R. & J. L. Santos, 584 (3).
Nascimento, O. C., 290 (4).
Neill, D. et al., 3934 (2); 6338 (2); 6485 (2).
Oldeman, R. A. A., 1111 (4); 2656 (4); B2992 (4).
Oliva, F., s.n. (2).
Oliveira, E., 53 (4); 1400 (suspected hybrid 2 × 4); 2267 (4); 2647 (4); 2838 (4); 3555 (4); 3970 (4); 4309 (4).
Øllgaard, B. et al., 35480 (2).
Omawale & R. Persaud, 122 (4); 216 (4).
Onishi, E., A. Ferreira & Taituba, 926 (4).
Ortega U., A. T., 127 (2).
Ortíz, F., 499 (4).
Ortíz V., P., 4287 (4).
Outer, R. W. den, 854 (4).
Palácios, W., 558 (2).
Pantin, L., s.n. (2).
Parker, C. S., s.n. (4).
Paulhinho Filho, H. F. & M. G. Silva, 83-82 (4); 83-156 (4).
Pavón y Jiménez, J. A., s.n. (4).
Pereira, E., 3346 (2).
Pérez-Arbeláez, E., s.n. (2); 687 (2); 10336 (2).
Perrottet, G., s.n. (4); 1382 (4).
Persaud, A. C., 30 (4); 50 (4).
Peters, C., 059/84 (2).
Philipson, W. R., J. M. Idrobo & A. Fernández, 1950 (2).
Pipoly, J., 11366 (4).
Pires, J. M., 927 (4); 1903 (4); 3705 (4); 5626 (4); 7397 (4); 51955 (4).
Pires, J. M. & P. B. Cavalcante, 51984 (4); 52680 (4).
Pires, J. M., W. Rodrigues & G. Irvine, 51157 (4); 51268 (4); 51425 (4).
Pires, J. M. & N. T. Silva, 11508 (2).
Pires, J. M. et al., 14055 (5); 16854 (2).
Pittier, H., 3785 (2); 3916 (2); 3980 (2); 4955 (2); 5459 (2); 8572 (2); 10956 (2).
Plowman, T., 2054 (2).
Plowman, T. & J. Schunke V., 7523 (2).
Plowman, T., N. A. Rosa & C. S. Rosário, 9855 (4).
Poeppig, E., s.n. (4); s.n. (2).
Poiteau, P. A., s.n. (4).
Poveda, L. J., 505 (2).
Prance, G. T. & T. D. Pennington, 1769 (4).
Prance, G. T., T. D. Pennington & N. T. Silva, 1306 (4).
Prance, G. T. & N. T. Silva, 58857 (2).
Prance, G. T. et al., 2161 (3); 2218 (4); 2483 (1); 3401 (5); 3797 (4); 3975 (2); 5229 (4); 5818 (5); 5992 (5); 6160 (4); 6379 (2); 6509 (4); 7716 (2); 7935 (2); 8381 (4); 8596 (5); 8819 (5); 10086 (4); 10870 (2); 12666 (2); 12980 (1); 13801 (5); 14140 (4); 14913 (3); 15157 (4); 16626 (3); 19451 (5); 20719 (4); 20753 (4); 20815 (4); 24944 (6); 55573 (2).
Projeto Flora, INPA no. 57110 (3).
Quiñones, L. M., 26 (2).
Rabelo, B. V., 297 (4); 333 (2); 1595 (4); 2282 (4).
Ramos, J., 120 (4).
Ramos, J. & G. Mota, 209 (4).
Ratter, J. R. et al., 1421 (4).
Renner, S. S., 1 (4); 8 (3); 188 (4); 337 (5); 365 (4); 366 (4); 502 (2); 533 (5); 594 (4); 786 (4); 922 (3); 929 (4); 945 (3); 958 (4); 959 (5); 968 (4); 970 (4); 971 (4); 983 (3); 986 (4); 987 (5); 990 (2); 1014 (4).
Rentería, E., F. Llanos & N. Sierra, 035 (4); 037 (2).

Rentería, E. et al., 1489 (2); 1491 (4); 1703 (2); 2000 (4); 2272 (2); 2343 (4).
Revilla, J. & J. Macedo, 1612 (2).
Reynel, C., 152 (2); 578 (2).
Ribeiro, B. G. S., 1139 (4).
Ribeiro, B. G. S. & G. S. Pinheiro, 1211 (4).
Richard, L. C. M., s.n. (4).
Riedel, L., 1480 (4).
Rimachi Y., M., 447 (2); 757 (2); 1783 (2); 2177 (2); 4424 (2); 5944 (2).
Rodrigues, I. A. et al., 857 (4).
Rodrigues, W., 214 (3); 2208 (4); 2848 (4); 3020 (4); 8927 (4).
Rodrigues, W. & J. Chagas, 1206 (3); 2440 (3); 2554 (4); 3572 (3); 4685 (4).
Rodrigues, W. & J. Lima, 2334 (4); 2750 (4); 3020 (4).
Rohl, P., 18 (4).
Romero-Castañeda, R., 1503 (2); 6495 (4); 8331 (2); 8346 (4); 8458 (4); 8460 (2).
Rosa, N. A. & M. R. Cordeiro, 1499 (4).
Rosemiro, 39 (4).
Roubik, D., 146 (4).
Ruiz-Terán, L., 1810 (2); 3175 (2).
Ruiz-Terán, L., A. Caralot & A. Mordes M., 10695 (2).
Ruiz-Terán, L. & S. Lopez-Palácios, 10963 (4).
Rusby, H. H., 722 (1); 2314 (2); 2330 (1).
Rutkis, E., 228 (4); 381 (2).
Saavedra S., H., 6337 (2).
Saer d'Heguert, J., 867 (4).
Sagot, P. A., s.n. (4); 238 (4).
Samuels, J. A., 519 (4).
Sandeman, C., 5583 (4); 5591 (2).
Saravía, C., 2728 (4).
Sastre, C., 1806 (4); 2008 (suspected hybrid 2 × 4); 3183 (4).
Saunders, J., 1075B (2).
Schipp, W. A., 319 (2).
Schnell, C., 158 (4); 404 (2).
Schultes, R. E., 3400 (2); 3926 (4); 19885 (4).
Schultes, R. E. & G. A. Black, 46-7 (2).
Schultes, R. E. & I. Cabrera, 13696 (4); 14922 (4); 17320 (4); 19248 (4).
Schultes, R. E., R. F. Raffauf & D. D. Saejarto, 26068 (4).
Schultes, R. E. & M. Villareal, 7364 (2).
Schunke, C., 601 (2).
Schunke V., J. M., 156 (2); 2237 (2); 2908 (1); 6582 (2).
Schwabe, W., 69-336 (4).
Schwacke, C. A. W., s.n. = Glaziou 9816 (3); III.n.374 = Glaziou 13830 (3).
Shattuck, O., 1119 (4).
Shepherd, J. D., 310 (2); 503 (4).
Silva, A. S. L., N. A. Rosa & R. P. Bahia, 1592 (4).
Silva, J. F., 128 (4); 474 (5).
Silva, Maria N., 335 (4).
Silva, Marlene F. et al., 798 (4); 1422 (2).
Silva, M. G., 1343 (3); 1510 (3); 1583 (3); 1780 (4); 1885 (4); 1887 (3); 2075 (2); 2168 (4); 4036 (3); 7117 (4).
Silva, M. G. & R. P. Bahia, 3562 (4).
Silva, M. G. & A. Pinheiro, 4299 (4).
Silva, M. G. & C. Rosário, 3685 (3).
Silva, M. G. & R. Souza, 2209 (3); 2220 (4).
Silva, N. T., 378 (5); 403 (5); 1745 (4); 1880 (3); 1883 (4); 2022 (3); 2081 (4); 3331 (4); 4336 (5).
Silva, N. T. & U. Brazão, 60801 (4).
Silva, N. T. & M. R. Santo, 4593 (4).
Simpson, D. R., 708 (2).
Skutch, A. F., 2441 (2).
Smith, A. C., 2818 (4).
Smith, D., 2889 (2).
Smith, D. et al., 1446 (4).
Smith Y., R. Z., 5153 (2).
Sneidern, K. von, 1334 (2).
Snethlage, E., MG no. 10054 (3).
Soejarto, D. D., 611 (2).
Soejarto, D. D. & J. D. Villa, 2650 (2).
Soejarto, D. D. et al., 1228 (4).
Solomon, J. C., 6387 (4); 7702 (4); 3238 (2).
Sparre, B., 18321 (2).
Sperling, C. R., 5749 (4).
Sperling, C. R. et al., 6232 (4).
Sprague, T. A., 391 (2).
Spruce, R., s.n. (3); 807 (3); 819 (3); 1145 (3); 1457 (3); 2162 (4); 3212 (2); 3268 (2).
Stahel Woodherb. Sur., 214 (4).
Standley, P. C., 25063 (2); 72336 (2); 72524 (2).
Stehlé, H., 955 (4); 3024 (4).
Stergios, B. & G. Aymard, 4031 (4).
Stern, W. L. et al., 261 (2); 672 (2).
Stevens, W. D., 8084 (2); 8475 (2); 12585 (2); 17475 (2).
Steyermark, J. A., 38084 (4); 39476 (2); 44533 (4); 44943 (4); 60588 (4); 86867 (4); 87386 (2); 99969 (2).
Steyermark, J. A. & G. Bunting, 102794 (4).
Steyermark, J. A., G. Bunting & C. Blanco, 101316 (2).
Steyermark, J. A. & V. C. Espinosa, 111657 (4).
Steyermark, J. A. & R. L. Liesner, 119221 (2).
Steyermark, J. A., R. L. Liesner & A. González, 120313 (2); 120549 (4).
Steyermark, J. A., R. L. Liesner & M. Guariglia, 127267 (4).
Steyermark, J. A. & J. L. Luteyn, 129711 (suspected hybrid 2 × 4).
Steyermark, J. A. et al., 117639 (4); 117670 (4).
Stoffers, A. L. et al., 321 (4).
Stork, H. E., 3084 (2).
Sullivan, G. A., 88 (2).
Teixeira, L. O. A. et al., 101 (4); 167 (5); 811 (4); 889 (5); 1360 (4).
Tessmann, G., 3523 (2).
Thomas, W. W., 3352 (4); 3375 (suspected hybrid 2 × 4).
Thomas, W. W., J. Guedes & R. P. Lima, 4056 (4); 4165 (6).
Thomas, W. W. et al., 3985 (4).
Tillett, S. S. & C. L. Tillett, 45287 (4); 45491 (4).
Todzia, C. et al., 2239 (4); 2264 (3).
Triana, J., s.n. (2); s.n. (4); 2 (4); 6288/1 (2).
Tonduz, A., 3916 = Pittier 3916 (2).
Trujillo, B., 13883 (2).
Türckheim, H. von, 7864 (4); 11242 (4).

Tunqui, S., 448 (2); 599 (2); 676 (2).
Tyson, E. L., 1119 (4).
Tyson, E. L. et al., 4735 (4).
Ubiratan, J., F. Ramos & C. D. Mota, 260 (4).
Uhl, C. F., 54 (4); 218 (4).
Ule, E., 5085 (3); 5664 (1); 5665 (2); 6097 (5); 8943 (4).
Uribe Uribe, L., 854 (4); 1339 (4); 2860 (2); 3045 (4); 4395 (4).
Utley, J. & K. Utely, 4923 (2).
Valério, M., 901 (2).
Vásquez, R. & N. Jaramillo, 1599 (2).
Vecht, J. van der, s.n. (4).
Veillon, J. P., 113 (4).
Vieira, G. et al., 62 (5); 445 (2).
Villwock, W., s.n. (2).
Vincelli, P., 1243 (4).
Vogel, S., 153 (2).
Wachenheim, H., 25 (4); 89 (4); 218 (4); 256 (4).
Weberbauer, A., 4505 (2).
Webster, G. L., K. Miller & L. Miller, 12412 (2).
Wedel, H. von, 2376 (2).
Wendt, T. et al., 2836 (2).
Werff, H. van der & A. González, 4882 (2).
Werner, WCAB 23/80PA (4).
White, G. & P. White, 140 (2).
Whiteford, C., s.n., 30 Jul 1976 (4).
Williams, L., 149 (2); 643 (2); 1588 (4); 2073 (2); 3969 (2); 4410 (2); 5591 (2); 7828 (2); 11886 (4); 14199 (7); 14342 (4); 14387 (7); 14652 (4); 15072 (4); 15088 (4); 15770 (2); 15771 (2).
Williams, L. O., A. Jimenez M. & T. P. Williams, 24301 (2).
Williams, L. O. et al., 28740 (2).
Williams, R. S., 576 (1); 808 (2).
Wilson-Browne, 546 = FD 5945 (4); 612 = FD 5979 (4).
Wolfe, F., 7A (2).
Woronow, G. & S. Juzepczuk, 4530 (4).
Woytkowski, F., 6040 (2).
Wullschlägel, H. R., 974 (4).
Wurdack, J. J. & L. S. Adderley, 42838 (4); 43080 (2); 43359 (4).
Wurdack, J. J. et al., 2813 (2).
Zarucchi, J. L., 1589 (4); 4875 (4).
Zarucchi, J. L. & R. Cárdenas, 4193 (2).
Zuluaga R., S., 171 (2).

NUMERICAL LIST OF TAXA: *LOREYA*

1. *L. nigricans* (Hooker f.) Triana
2. *L. wurdackiana* Renner
3. *L. ovata* O. Berg ex Triana
4. *L. arborescens* (Aublet) DC.
5. *L. umbellata* (Gleason) Wurdack
6. *L. gracilis* Renner
7. *L. subrotundifolia* (Wurdack) Renner
8. *L. riparia* Renner
9. *L. strigosa* Gleason
10. *L. subandina* Wurdack
11. *L. mespiloides* Miquel
12. *L. spruceana* Bentham ex Triana
13. *L. klugii* Renner

LIST OF EXSICCATAE: *LOREYA*

Numbers in parentheses following each collection refer to the species numbered above.

Acosta Solis, M., 13742 (5).
Alencar, I. L., 599 (2).
Allard, H. A., 22575 (12).
Ancuash, E., 16 (10), 434 (13), 1123 (12).
Araquistian, 3044 (11), 3221 (11).
Asplund, E., 9181 (10), 9463 (10).
Aublet, J. B. C. F., s.n. (4).
Aymard, G. et al., 3919 (11).
BAFOG, 1199 (4), 1237 (4), 3793 (4), 4307 (11), 7210 (11), 7282 (11).
Bailey, I. W., 100 (11).
Barrier, S., 585 (13).
Baumrucker, J. & M. Calzadilla, s.n. (12).
Beck, S. G., 1488 (13).
Benoist, R., 184 (11), 1588 (4).
Bernardi, L., 2104 (12).
Berry, P., O. Huber & P. Redmond, 113894 (11).
Billiet, F. & B. Jadin, 1560 (11), 1714 (11).
Boeke, J. D. & Ramirez, 1264 (10).
Boldingh, I., 3918 (11).
Boom, B. M. & A. L. Weitzman, 5441 (4).
Boom, B. M. et al., 5365 (3), 5707 (12).
Breteler, F. J., 4046 (12).
Buchtien, O., 1034 (10).
B.W., 516, 1176, 2355, 2459, 2510 = 1578, 2586, 2913, 3316, 3459, 3887, 4186, 4390 = 634, 4707, 4774, 4989, 6582 (11).
Cardenas, M., 1199 (12).
Castillo, A., 2052 (11).
Cazalet, P. C. D. & T. D. Pennington, 5049 (5), 7724 (12).
Cavalcante, P., 2527 (11).
Chacon, 1950 (11).
Clark, H. L., 6866 (4), 7086 (3).
Cooper, P., 594 (11).
Cremers, G., 6597 (11), 7652 (11).
Croat, T. B., 18533 (5), 18716 (5), 54198 (11).
Croizat, L., 742 (11).
Cuadros, H. & A. Gentry, 2979 (12).
Cuatrecasas, J., 8917 (9).

Davidse, G. & J. S. Miller, 26540, 27483 (12).
Davis, E. W., 153 (5), 370 (12).
Davis, E. W. & J. Yost, 937 (12).
Dawson, 13143 (11).
De La Cruz, J. S., 4426 (11), 4565 (11).
Delascio, F. & R. L. Liesner, 7281 (11).
Diaz, M., 64A (4), 152A (5).
Dodson, C. H., 2807 (5), 5826 (5), 7410 (5).
Dodson, C. H. & J. A. Duke, 7641 (5).
Donselaar, J. van, 3078 (11), 3451 (11).
Ducke, A., s.n. = RB25507 (2), s.n. = RB 41350 (4), 171 (1), 274 (1), 1464 (12), 1720 (4), 1722 (4).
Dwyer, J. et al., 4366 (11).
Encarnación, F., 26250 (4).
FD, 2730 = 121 (11), 3391 = 655 (4), 3314 = 578 (4), 4522 = 1786 (4).
Fernández, A., 969 (11), 1985 (3).
Fernández Casas, J. & Susanna, 8255 (12).
Feuillet, C., 354 (11).
Focke, H. C., 946 (11).
Foresta, H. de, 637 (4).
Freitas, C. A. A. et al., 120 (12).
Fróes, R. L., 20513 (1), 34656 (4).
García-Barriga, H., 14066 (4), 14420 (5).
Gentry, A. & G. Shupp, 26673 (5).
Gentry, A. et al., 22194 (12), 22430 (5), 24994 (5), 25429 (12).
Gieteling, C. J., 21 (11).
Gleason, H. A., 691 (11).
Granville, J.-J., 2815 (7), 2999 (11), B-3680 (11), 3897 (11), 4489 (7), 4629 (7).
Granville, J.-J. et al., 6312 (11).
Hammel, B., 11652 (11), 14121 (11).
Harling, G. & L. Andersson, 11968 (10).
Haxaire, G., 937 (13).
Hekking, W. H. A., 860 (11), 1202 (11).
Hohenkerk, L. S., 874 (4).
Holm-Nielsen, L. et al., 21748 (12).
Holst, B. & R. Liesner, 3414 (4).
Huashikat, V., 56 (10), 1990 (12).
Irvine, D., 11 (12), 128 (12), 192 (12), 408 (12).
Jimenez-Saa, H., 1504 (11).
Jong, B. de, 2 (7).
Kappler, A., 1461a (11).
Kayap, R., 1072 (12).
Killip, E. P. & A. C. Smith, 27455 (10).
Klug, G., 19 (5), 131 (5), 709 (5), 859 (5), 1927 (13), 1968 (13), 2545 (5).
Kramer, K. U. & W. H. A. Hekking, 2515 (11).
Krukoff, B. A., 1510 (9), 1593 (9), 8708 (3), 11221 (6), 12290 (11).
Kuhlman, M. & S. Jimbo, 226 (4).
Kujikat, A., 293 (12).
Kuyper, J., 14 (11).
Lanjouw, J. & J. C. Lindeman, 976 (11), 1176 (11).
Lasser, T., 1648 (11).
LBB, 13143 (11), 14245 (11).
Lent, R. W., 1517 (11).
Leveau, J. A., 37 (10).
Liesner, R. L., 3715 (4), 7383 (3), 19124 (11).
Lindeman, J. C., 79 (11), 674 (11), 4587 (11), 5064 (11), 6145 (11), 6448 (11).
Linder, D. H., 90 (11).
Little, E. L., s.n. 30 Sep 1975 (11), 584 (10), 6223 (5).
Lowrie, S. R. et al., 587 (9).
Maguire, B. & D. B. Fanshawe, 22967 (12).
Maguire, B. & C. K. Maguire, 40176 (11).
Maguire, B. & L. Politi, 28164 (4).
Maguire, B., J. J. Wurdack & W. M. Keith, 41778 (4).
Maia, L. A., 599 (2).
Marcano-Berti, L., L. A. Pinto & I. Peña S., 50981 (4).
McDaniel, S. & M. Rimachi Y., 17210 (13), 26313 (5).
Mélinon, C., 147 (11), 357 (11).
Mejía, K., 320 (5).
Mennega, A. M. W., 337 (11).
Mori, S. & A. Bolten, 8384 (11).
Mori, S. & C. A. Gracie, 18327 (11).
Mori, S., B. M. Boom & Keller, 14907 (11).
Nascimento, O. C., 761 (2).
Nee, M. & Vega, 27856 (11).
Nee, M. & J. C. Solomon, 30308 (6).
Neill, D. et al., 6550 (10).
Nelson, B. W. & S. Nelson, 1444 (8).
Oldeman, R. A. A., B-2041 (11), 2266 (11), 2755 (11).
Øllgaard, B. et al., 34595 (10).
Outer, R. W. den, 965 (11).
Palácios, W., D. Neill & J. Zaruma, 00414 (12).
Pérez-Arbeláez, E., 753 (9).
Pires, J. M., 214, 225 (1), 904 (4), 7003 (4), 50299 (cf. 11, see note under that species).
Poiteau, P. A., s.n. (4).
Poole, J. M., 2011 (2).
Prance, G. T. & T. D. Pennington, 1839 (4).
Prance, G. T. et al., 4699 (12), 8868 (9), 10275 (11), 11440 (12), 13168 (9), 16494 (9), 20301 (12), 23439 (5), 23941 (3).
Prevost, M. F., 823 (11).
Proctor, J., 4765 (11).
Rabelo, B. V. & R. Cardoso, 2910 (11).
Rabelo, B. V., N. Solaman & R. Souza, 3048 (7).
Ramírez, C. R., 094 (5).
Renner, S. S., 131 (12), 237 (12), 597 (12), 930 (8), 963 (12), 1012 (12), 1015 (8).
Revilla, J., 1295 (5).
Rimachi, Y. M., 1872, 3123, 3625, 3749, 6278 (5).
Rosa, N. A. & O. C. Nascimento, s.n. = 3357 (12).
Ruíz & Pavón, 16/59 (12).
Ruiz-Terán, L. & S. López-Palácios, 11104 (11).
Sagot, P. A., 239 (11).
Salazar, A., 022 (10).
Samuels, J. A., s.n. 12 Apr 1916 (11), 271 (11).
Sandino, 4783 (11).
Santos, M. R., 225 (11), 1533 (12).
Sastre, C., 6017 (11).
Schnell, C., 11290 (11), 11365 (11).
Schultes, R. E., 3629 (11), 9033 (4).
Schultes, R. E. & I. Cabrera, 14038 (4), 15659 (3), 15916, 16628, 17094, 17349 (5), 17913 (4), 19755 (4).
Schultes, R. E., R. E. D. Baker & I. Cabrera, 17946 (4).
Schultes, R. E. & F. López, 9360 (4).
Schultes, R. E., R. F. Raffauf & D. Soejarto, 24417 (4).
Schunke V., J. M., 8111 (12), 10093 (12).
SEF, 10178 (10).
Silva, A. S. L. da, N. A. Rosa & R. P. Bahia, 404 (5).

Silva, N. T. & U. Brazão, 60742 (12), 60914 (4).
Skog, L. et al., 7042 (11).
Smith, D. N., 4012 (12).
Smith, E. E. & J. Vera, 345 (12).
Spruce, R., s.n. (1), 1144 (12), 1249 (12), 1331 (1), 2588 (4), 2899 (3), 3085 (4).
Stahel Woodherb. Sur., 306 (11).
Steyermark, J. A., 75591 (11), 90367, 90693, 96690, 99940 (12).
Steyermark, J. A. & M. Rabe, 96690 (12).
Steyermark, J. A. et al., 113894 (11).
Teixeira, L. O. A. et al., 365 (12).
Thomas, W. W., 3328 (12).
Tillett, S. S. & C. L. Tillett, 45763 (11), 672-110 (10).
Traill, J. W. H., 311 (3).
Tunqui, S., 1016 (12), 1051 (12).
Ule, E., 5344 (1).
Vásquez, R. et al., 628 (5), 1299 (12).
Vásquez, R. & N. Jaramillo, 4482, 6778 (5), 7959.
Viera, G. et al., 472, 721 (11).
Wachenheim, H., 21 (11).
Wessels-Boer, J. G., 1385 (11).
Williams, L., 13919, 14186, 15139, 15639 (3).
Williams, R. S., 662 (10).
WWF Minimal Critical Size Project, 1103-424 (12), 1103-453 (12).
Zamora V., N. & Q. Jiménez M., 1285 (11).
Zarucchi, J. L., 2358 (10).

Numerical List of Taxa: *Macairea*

1. *Macairea duidae* Gleason
2. *M. chimantensis* Wurdack
3. *M. thyrsiflora* De Candolle
4. *M. theresiae* Cogniaux
5. *M. spruceana* Berg ex Triana
6. *M. sulcata* Triana
7. *M. rufescens* De Candolle
8. *M. maroana* Wurdack
9. *M. cuieirasii* Renner
10. *M. pachyphylla* Bentham
11. *M. radula* (Bonpland) De Candolle
12. *M. parvifolia* Bentham
13. *M. rigida* Bentham
14. *M. neblinae* Wurdack
15. *M. linearis* Gleason
16. *M. cardonae* Wurdack
17. *M. lanata* Gleason
18. *M. philipsonii* Renner
19. *M. multinervia* Bentham
20. *M. lasiophylla* (Bentham) Wurdack
21. *M. stylosa* Triana
22. *M. axilliflora* Wurdack

List of Exsiccatae: *Macairea*

Numbers in parentheses following each collection refer to the species numbered above.

Aguirre-Galviz, L. E., 1070 (6).
Alencar, L., 623 (3).
Allen, P. H., 3193 (3).
Almeida, E. de F., 170 (11).
Almeida, S. P. et al., 131 (11).
Alvarenga, M., s.n. = RJ 90559 (3).
Amarál, D. L., 119 (11).
Amarál, I. L., 1680 (3); 1626 (10).
Amarál, I. L. et al., 1014 (11); 1107 (10).
Anderson, A. B., 138, 182 (4).
Anderson, W. R., 9644, 9724 (11); 10766, 10921 (3); 13381 (14).
Andrade, A. G. & M. Emmerich, 968 (11).
Araque Molina, J. & F. A. Barkley, 18V202 (3).
Argent, G. et al., 6318 (11).
Arrojado Lisbôa, M., 2344 (11).
Assis, V., 1 (11).
Bahia, T. R., 29 (3).
Barclay, A. S. et al., 522 (3).
Barros, F. de, 896 (11).
Beck, S. G. & R. Haase, 10072 (3).
Becker, J., s.n. = 142024 (11).
Bello, W., 1075 = 140 (11).
Berg, O. K., s.n. (11).
Bernardi, L., 1688 (10); 2631 (12); 6613 (10).
Black, G. A., 48-2590 (5); 55-18527 (3).
Black, G. A. & P. Ledoux, 50-10569 (3).
Black, G. A. et al., 57-19510 (3).
Blanchet, J. S., 2900 (11).
Bockermann, W., 171, 222 (3).
Brade, A. C., s.n. = RJ 10559 (5); 13760, 13977 (11).
Brewer-Carías, C., 27 Dec 1977 s.n. (11).
Brown, B., No. A (20).
Bunting, G. S. et al., 3602 (3); 4013 (8).
Burchell, W. J., 5377, 5781 or 4, 7939 (11).
Calderon, C. E. et al., 2556, 2677, 2723, 2821 (3).
Campos Porto, P., 1206 (11).
Cardona, F., 256 (16); 930, 958 (12); 1555 (2); 1840, 1841, 1861 (10); 1862 (12); 1866 (16); 1872 (12); 1894 (10); 1961 (2); 2022, 2065, 2233 (10); 2283 (16); 2302 (2); 2426 (3); 2427 (2); 2468 (12); 2617 (10); 2652 (2); 3021 (12).
Cid, C. A. et al., 238, 465, 2295, 2310 (4); 2481 (3); 4410 (10); 5516 (19); 5898 (3).
Clark, H. L., 6467, 6519 (7); 7506 (22); 7517 (21).
Clark, H. L. & P. Maquirino, 8230, 8234 (21); 8235 (3).
Claussen, P., s.n., 26, 49, 131, 344A, 345, 569, 786, 1026 (11).
Cobra, L. Q., 298, 600 (11).
Coêlho, D., s.n. = INPA 3999, s.n. = INPA 4046, s.n. = INPA 20601 (4).

Kral, R., 70487 (20); 71865 (14); 72135 (12).
Kral, R. & A. C. Gonzalez, 70484 (19); 70503 (3).
Kramer, K. U. & W. H. A. Hekking, 2933 (10).
Krapovickas, A. & A. Schinini, 31991 (11); 35065 (3).
Kuhlmann, M., s.n. = May 1918 (10); 3613, 4150 (11).
Kuntze, O., s.n. (11).
Labauriau, L., s.n. = RJ 68274 (11).
Landsberge, van, 15 (3); 17, 18 (21); 25 (3).
Lanstyack, L., s.n. = RB 55995 (11).
Lasser, T., 1960 (19).
Liesner, R. L., 3408, 3859 (7); 4041 (3); 6867 (7); 16090 (14); 17194 (3); 18162 (1); 19099 (10); 19307 (12); 19360, 20694 (20).
Liesner, R. L. & H. Clark, 9121 (22).
Liesner, R. L. & B. Holst, 18792 (19); 18809 (10); 21426 (13).
Liesner, R. L. & G. Morillo, 18615 (1).
Liesner, R. L. et al., 21004, 21050 (2).
Lima, A. S., 61-3832, 5300, 6274 (11).
Lima, J., 461 (3).
Löfgren, A., 806, 1145 (11).
Löfgren, A. & G. Edwall, 2028 (11).
Loureiro, A. et al., s.n. = INPA 43837 (4).
Lützelburg, P. v., 1280, 12226, 16048 (11); 21511 (20); 22465, 22472 (3).
Lund, P. W., s.n. (11).
Luteyn, J. L. et al., 6326 (12); 9457 (14); 9491 (2); 9524 (16).
Maas, P. J. M. & O. Huber, 5121 (3); 5372 (10).
Maas, P. J. M. et al., 4176 (10); 5765 (3); 5823 (10).
Macedo, A., 33, 422, 447, 1163, 3263, 4459, 4631 (11).
Maciel, A., 79 (3).
Maciel, A. & A. Webber, 110 (11).
Maguire, B., 24350, 24793 (10); 29251 (21); 32805, 32928 (12); 33259 (10); 33286 (3); 33587 (12); 33670 (19); 33673 (20).
Maguire, B. & B., 29105, 29166 (13).
Maguire, B. & C. K., 35068, 35211 (13); 35418 (10); 35451 (13); 35474, 40344 (10); 40360, 40378 (20).
Maguire, B. & D. Fanshawe, 23110 (3); 23342 (10).
Maguire, B. & L. Politi, 28324 (13).
Maguire, B. & J. Wurdack, 33792 (12); 33824 (3); 33843 (12); 33989 (10); 33995 (19); 34503 (3); 34599 (21); 34680, 34874, 35559 (3); 35560 (17); 35605 (21); 35611 (5); 35675 (17); 35724 (22).
Maguire, B. et al., 29552, 29644 (1); 29725 (13); 30510, 30527, 30555, 30574, 30668, 30866 (3); 30890 (17); 30891, 30919 (3); 30925, 31685 (13); 36348 (17); 36409 (7); 36429 (8); 36440 (17); 36495 (21); 36562, 36613, 36633 (3); 36849, 36935, 37198, 37212 (14); 37567 (3); 37600 (21); 37601 (22); 37610 (5); 37638 (3); 37668 (22); 37692, 40696, 41466, 41467 (3); 41482 (21); 41510, 41573 (3); 41709, 41744 (7); 41772 (8); 41800 (3); 41832 (5); 41847 (17); 41882 (3); 41904 (17); 42095, 42101, 42423 (14); 42633 (3); 42633 (22); 43866 (10); 44115 (3); 44157 (7); 44168 (3); 46190A (20); 46217 (10); 53663 (3); 56129, 56185, 56221, 56325 (11).
Malme, G. O. A., s.n., 1610B, 1693 (11); 3486 (3).
Manso, A. L. P. da Silva, 177 (11).
Marcano-Berti, L. & P. Salcedo, 141-979 (7).
Marcano-Berti, L. et al., 128-981 (12).
Markgraf, F. et al., 3356, 3516 (11).
Martinelli, G., 6882 (3).
Martins, F. R., 3562 (11).
Martius, C. F. P. v., s.n., 837 (11); 3156 (7).
Mattos, A. & E. P. Heringer, 352 (11).
McConnell, F. V. & J. J. Quelch, 31 (3).
McDaniel, S., 16457 (3).
Meader, E., 15 (3).
Medina, E., s.n. (7); 554 (5).
Mello Barreto, H. L. de, 6866, 6869, 6870, 6871, 6872, 9278, 9348 (11).
Meyer, H., 634 (11).
Miranda, C. A., 322 (11).
Moore, H. E., 9674 (19); 9679 (20).
Moore, H. E. et al., 9702 (3).
Mori, S. A. et al., 16822 (11).
Morillo, G., 4167, 4999, 5202 (7); 5292 (3); 5375 (7); 6922 (3).
Morillo, G. et al., 3892 (7).
Mosén, H., 1287 (11).
Mota, C. D. A., s.n. = INPA 60485, s.n. = INPA 60469 (3).
Mota, C. D. A. & O. P. Monteiro, s.n. = INPA 61005 (3).
Nascimento, O. C., 769 (3).
Nestor, J., s.n. (11).
Novaes, J. C. de Campos, 567 (11).
Oldenburger, F. et al., 184, 963, 1418, 1421 (10).
Oliveira, J. E. de, 1067 (11).
Orbigny, C. V. D. d', 687 (11).
Pabón E., M. et al., 423 (17).
Pannier, F. & W. Schwabe, s.n. = VEN 46308 (3), s.n. (2).
Pereira, B. A. S., 551, 620 (11).
Pereira, E., 1672 (11).
Pereira, E. & A. P. Duarte, 10376 (11).
Phelps, K. D., 399 (3).
Phelps, K. D. & C. B. Hitchcock, 470 (17); 510, 523 (13).
Philipson, W. R., 2378 (18).
Pipoly, J. J. & G. Samuels, 6840 (3).
Pipoly, J. J. et al., 7205 (2).
Pires, J. M., 16070 (3); 17364 (20).
Pires, J. M. & P. Leite, 14538 (4).
Pires, J. M. & L. R. Marinho, 15709 (8).
Pires, J. M. & A. Mattos, 9866 (11).
Pires, J. M. & M. R. Santos, 16197, 16405, 16501 (11).
Pires, J. M. & N. T. Silva, 4722 (4).
Pires, J. M. et al., 14215 (9).
Plowman, T., 13714 (7).
Pohl, J. E., 298, 620, 1260, 1296, 1313 (11).
Poole, J. M., 1875 (3).
Prance, G. T., 5680 (10); 15537 (21).
Prance, G. T. & O. Huber, 28434, 28458 (3).
Prance, G. T. & E. Lleras, 23732 (4).
Prance, G. T. & N. T. Silva, 58466 (11).
Prance, G. T. et al., 1288 (3); 5559, 5834, 5983, 9900 (10); 13775 (3); 14837 (4); 15462 (21); 17931, 21080, 21103, 23485 (4); 25244 (3); 29167A (12); 29167B (3); 29257 (10); 29814 (3).
Ramirez, N., 740 (12); 944 (10).
Ramos, J. et al., s.n. = INPA 62113 (3).
Ratter, J. A., 2000 (3); 3108 (11).
Ratter, J. A. & R. A. de Castro, 166 (11).

Index of Local Names

INDEX TO SCIENTIFIC NAMES

New names and combinations are in **boldface** and synonyms are in *italics*. Page numbers in **boldface** indicate primary page references. Page numbers with an asterisk (*) indicate pages with illustrations or maps.

Susanne S. Renner

Dr. Susanne S. Renner was born in Tübingen, West Germany in 1954. She attended the University of Hamburg, where she received her Diploma in Biology in 1980 and the Dr. rer. nat. degree in 1984. After fieldwork in Brazil (1980–1982 and 1984), she was a postdoctoral fellow at the Smithsonian Institution in Washington, D.C. (1984–1987). Dr. Renner has studied and collected tropical plants in Brazil, Venezuela, and Ecuador. Her research has focused primarily on the taxonomy and ecology of New World Melastomataceae. At present, she is associate professor at the Botanical Institute of the University of Aarhus, Denmark.

MEMOIRS OF THE NEW YORK BOTANICAL GARDEN

The Botany of the Guayana Highland—Parts I–XIII
by Bassett Maguire and Collaborators

I.	Mem. N. Y. Bot. Gard. 8(2): 87–160. 1953. (Combretaceae, Compositae, Eriocaulaceae, Leguminosae, Malpighiaceae, Melastomataceae, Polygalaceae, Rutaceae)	$10.00
II.	Mem. N. Y. Bot. Gard. 9(3): 235–392. 1957. (Annonaceae, Begoniaceae, Bignoniaceae, Bromeliaceae, Compositae, Droseraceae, Eriocaulaceae, Gramineae, Haemadoraceae, Leguminosae, Musci, Piperaceae, Polygalaceae)	$15.00
III.	Mem. N. Y. Bot. Gard. 10(1): 1–156. 1958. (Apocynaceae, Bignoniaceae, Combretaceae, Guttiferae, Leguminosae, Melastomataceae, Myrtaceae, Rapateaceae, Xyridaceae)	$15.00
IV.	Mem. N. Y. Bot. Gard. 10(2): 1–37. 1960. (Musci)	$10.00
IV.	Mem. N. Y. Bot. Gard. 10(4): 1–87. 1960. (Acanthaceae, Annonaceae, Aquifoliaceae, Bromeliaceae, Guttiferae, Linaceae, Marantaceae, Melastomataceae, Ochnaceae, Rutaceae, Xyridaceae)	$10.00
V.	Mem. N. Y. Bot. Gard. 10(5): 1–278. 1960. (Bambusoideae, Bromeliaceae, Guttiferae, Lauraceae, Melastomataceae, Piperaceae, Rubiaceae, Xyridaceae)	$17.50
VI.	Mem. N. Y. Bot. Gard. 12(3): 1–285. 1965. (Aquifoliaceae, Cyperaceae, Euphorbiaceae, Gramineae, Lauraceae, Rapateaceae, Rubiaceae)	$17.50
VII.	Mem. N. Y. Bot. Gard. 17(1): 1–439. 1967. (Abolbodaceae, Bombacaceae, Compositae, Cyperaceae, Euphorbiaceae, Labiatae, Lentibulariaceae, Polypodiaceae, Rubiaceae)	$35.00
VIII.	Mem. N. Y. Bot. Gard. 18(2): 1–290. 1969. (Annonaceae, Apocynaceae, Bromeliaceae, Cyperaceae, Gramineae, Malpighiaceae, Melastomataceae, Myrtaceae, Quiinaceae, Schizaeaceae, Velloziaceae)	$20.00
IX.	Mem. N. Y. Bot. Gard. 23: 1–832. 1972. (Bonnetiaceae, Caryocaraceae, Clusiaceae, Cycadeaceae, Cyclanthaceae, Gramineae, Meliosmaceae, Melastomataceae, Palmae, Piperaceae, Polygalaceae, Polypodiaceae, Rubiaceae, Sapotaceae, Tetrameristaceae)	$50.00
X.	Mem. N. Y. Bot. Gard. 29: 1–288. 1978. (Bonnetiaceae, Bignoniaceae, Cyatheaceae, Ericaceae, Icacinaceae, Saccifoliaceae, Sarraceniaceae, Styracaceae, Symplocaceae, Tiliaceae)	$25.00
XI.	Mem. N. Y. Bot. Gard. 32: 1–391. 1981. (Tepuianthaceae, Malpighiaceae, Dipterocarpaceae, Convolvulaceae, Ebenaceae, Lissocarpaceae, Gentianaceae)	$40.00
XII.	Mem. N. Y. Bot. Gard. 38: 1–84. 1984. (Hymenophyllopsidaceae, Hymenophyllaceae, Araliaceae)	$16.00
XIII.	Mem. N. Y. Bot. Gard. 50: 1–112. 1989. (Gentianaceae, *Ouratea* [Ochnaceae], Melastomataceae, Dioscoreaceae, Nyctaginaceae, Monimiaceae, Capparaceae, Cunoniaceae, Rhamnaceae, Rutaceae)	$

Postage and Handling:

U.S. Orders: $2.00 + 3% of subtotal.
Non-U.S. Orders: $3.00 + 4% of subtotal. Payment in U.S. currency drawn on a U.S. bank. Thank you.

Order from: MEMOIRS, Scientific Publications Department, The New York Botanical Garden, Bronx, NY 10458 USA.